112
Structure and Bonding

Series Editor:
D.M.P. Mingos

Structure and Bonding

Recently Published and Forthcoming Volumes

Principles and Applications of Density Functional Theory in Inorganic Chemistry II
Volume Editors:
Kaltsoyannis, N., McGrady, J.E.
Vol. 113, 2004

Principles and Applications of Density Functional Theory in Inorganic Chemistry I
Volume Editors:
Kaltsoyannis, N., McGrady, J.E.
Vol. 112, 2004

Supramolecular Assembly via Hydrogen Bonds II
Volume Editor: Mingos, D.M.P.
Vol. 111, 2004

Applications of Evolutionary Computation in Chemistry
Volume Editor: Johnston, R.L.
Vol. 110, 2004

Fullerene-Based Materials
Structures and Properties
Volume Editor: Prassides, K.
Vol. 109, 2004

Supramolecular Assembly via Hydrogen Bonds I
Volume Editor: Mingos, D.M.P.
Vol. 108, 2004

Optical Spectra and Chemical Bonding in Transition Metal Complexes
Special Volume II
dedicated to Professor Jørgensen
Volume Editor: Schönherr, T.
Vol. 107, 2004

Optical Spectra and Chemical Bonding in Inorganic Compounds
Special Volume I
dedicated to Professor Jørgensen
Volume Editor: Schönherr, T.
Vol. 106, 2004

Group 13 Chemistry III
Industrial Applications
Volume Editors:
Roesky, H.W., Atwood, D.A.
Vol. 105, 2003

Group 13 Chemistry II
Biological Aspects of Aluminum
Volume Editors:
Roesky, H.W., Atwood, D.A.
Vol. 104, 2002

Group 13 Chemistry I
Fundamental New Developments
Volume Editors:
Roesky, H.W., Atwood, D.A.
Vol. 103, 2002

High Performance Non-Oxide Ceramics II
Volume Editor: Jansen, M.
Vol. 102, 2002

High Performance Non-Oxide Ceramics I
Volume Editor: Jansen, M.
Vol. 101, 2002

π-Electron Magnetism
From Molecules to Magnetic Materials
Volume Editor: Veciana, J.
Vol. 100, 2001

Principles and Applications of Density Functional Theory in Inorganic Chemistry I

Volume Editors:
N. Kaltsoyannis · J.E. McGrady

With contributions by
J. Autschbach · E.J. Baerends · O. Gritsenko · J.N. Harvey
F. Maseras · G. Ricciardi · A. Rosa · G. Ujaque

Springer

The series *Structure and Bonding* publishes critical reviews on topics of research concerned with chemical structure and bonding. The scope of the series spans the entire Periodic Table. It focuses attention on new and developing areas of modern structural and theoretical chemistry such as nanostructures, molecular electronics, designed molecular solids, surfaces, metal clusters and supramolecular structures. Physical and spectroscopic techniques used to determine, examine and model structures fall within the purview of Structure and Bonding to the extent that the focus is on the scientific results obtained and not on specialist information concerning the techniques themselves. Issues associated with the development of bonding models and generalizations that illuminate the reactivity pathways and rates of chemical processes are also relevant.
As a rule, contributions are specially commissioned. The editors and publishers will, however, always be pleased to receive suggestions and supplementary information. Papers are accepted for *Structure and Bonding* in English.
In references *Structure and Bonding* is abbreviated *Struct Bond* and is cited as a journal.

Springer WWW home page: http://www.springeronline.com
Visit the SB content at http://www.springerlink.com/

ISSN 0081-5993
ISBN 3-540-21860-2
DOI 10.1007/b12224
Springer Berlin Heidelberg New York

Library of Congress Control Number 2004104815

Springer is a part of Springer Science+Business Media

springeronline.com

Printed in Germany

Cover: KünkelLopka GmbH, Heidelberg/Design & Production GmbH, Heidelberg
Typesetting: Stürtz GmbH, Würzburg
Production Editor: Angélique Gcouta, Berlin

Printed on acid-free paper 02/3020 gc – 5 4 3 2 1 0

Series Editor

Volume Editors

Dr. Nikolas Kaltsoyannis
Chemistry Department
University College London
20 Gordon Street
London WC1H 0AJ, UK
uccanka@ucl.ac.uk

Dr. John E. McGrady
Department of Chemistry
Inorganic Chemistry
University of York
Heslington
York, YO 105DD, UK
jem15@york.ac.uk

Editorial Board

Structure and Bonding is also Available Electronically

For all customers who have a standing order to Structure and Bonding, we offer the electronic version via SpringerLink free of charge. Please contact your librarian who can receive a password for free access to the full articles by registering at:

http://www.springerlink.com

If you do not have a subscription, you can still view the tables of contents of the volumes and the abstract of each article by going to the SpringerLink Homepage, clicking on "Browse by Online Libraries", then "Chemical Sciences", and finally choose Structure and Bonding.

You will find information about the

- Editorial Board
- Aims and Scope
- Instructions for Authors
- Sample Contribution

at http://www.springeronline.com using the search function.

Preface

It is difficult to overestimate the impact that density functional theory has had on computational quantum chemistry over the last two decades. Indeed, this period has seen it grow from little more than a theoreticalcuriosity to become a central tool in the computational chemist's armoury. Arguably no area of chemistry has benefited more from the meteoric rise in density functional theory than inorganic chemistry. the ability to obtainreliable results in feasible timescales on systems containing heavy elements such as the d and f transition metals has led to an enormous growth in computational inorganic chemistry. The inorganic chemical literature reflects this growth; it is almost impossible to open a modern inorganic chemistry journal without finding several papers devoted exclusively or in part to density functional theory calculations. The real importance of the rise in density functional theory in inorganic chemistry is undoubtedly the much closer synergy between theory and experiment than was previously posible.

In these volumes, world-leading researchers describe recent developments in the density functional theory and its applications in modern inorganic and bioinorganic chemistry. These articles address key issues key issues in both solid-state and molecular inorganic chemistry, such as spectroscopy, mechanisms, catalysis, bonding and magnetism. The articles in volume I are more focussed on advances in density functional methodogy, while those in Volume II deal more with applications, although this is by no means a rigid distinction. We are very grateful to all our contributors for their extensive efforts. We are very proud to present this collection of world-class science, and hope that you enjoy reading it as much as we have enjoyed putting it together.

September 2004

Nikolas Kaltsoyannis
John E. McGrady

Contents

Contents of Volume 113

Principles and Applications of Density Functional Theory in Inorganic Chemistry II

Volume Editors: N. Kaltsoyannis · J.E. McGrady
ISBN 3-540-21861-0

Structure and Bonding (2004) 112:1–48
DOI 10.1007/b97936

The Calculation of NMR Parameters in Transition Metal Complexes

Jochen Autschbach (✉)

Department of Chemistry, University at Buffalo, Buffalo, NY 14260-3000, USA
jochena@buffalo.edu

Abstract An overview is presented of the methodology and computations of nuclear shielding and spin-spin coupling constants of transition-metal complexes. The material presented also includes an outline of relativistic approaches and their applications to heavy transition-metals.

Keywords Density functional calculations · Transition metal complexes · NMR chemical shifts · Nuclear spin-spin coupling constants · Relativistic quantum chemistry

1 Introduction

It is fair to state that reasonably accurate computations of NMR parameters for transition-metal complexes have only become feasible owing to the enormous development of density functional theory (DFT) methods for chemical applications (along with powerful computer hardware) during the 1990s. This has perhaps two main reasons: (1) the strong influence of electron correlation in particular for the 3*d* metals, which renders Hartree–Fock calculations of NMR parameters often unreliable (with exceptions, see Sect. 4), and (2) the large number of electrons and the need for a relativistic treatment for complexes of the 4*d* and 5*d* metals. Relativistic DFT methods have also only during the last decade become efficient and accurate enough for routine applications in transition-metal chemistry. This has in turn allowed the subsequent development of relativistic NMR program codes.

The application of DFT methods to the computation of transition-metal NMR has been reviewed in the past [1–4]. A short overview was recently prepared by Bühl [5]. NMR calculations on heavier transition-metal complexes have further been discussed in reviews devoted to relativistic NMR methodology [6–9]. Thus, the present overview does not attempt to give a full coverage of the available literature, but to present a number of illustrative examples, the present status of such computations and their accuracy and limitations, along with a description of the underlying methodology. Because of the high importance of relativistic effects on NMR parameters, which is clearly represented in the available literature on DFT NMR computations of transition-metal complexes, the reader will find that a substantial portion of this paper is devoted to this topic.

The methodological section is intended for readers who are not very familiar with the theoretical details. Accordingly, only part of the formalism is presented in the form of equations; however, the general quantum mechanical concepts that allow the computation of NMR observables from first-principles theory, as well as some important technical details, are outlined. Readers who are not interested in methodological details may skip Sects. 2.2 – 2.9, since the subsequent sections do not make excessive references to the methodology part.

The discussion of theory as well as computational results is restricted to closed-shell systems. An account of calculations of NMR parameters for paramagnetic systems can be found in Ref. [10]. Further, most of the applications that are discussed in Sect. 4 are not concerned with the calculation of vibrational corrections to NMR parameters [11]. Most methodolocial advances in this field have so far been tested on comparatively small molecules containing main-group elements. Only recently, some work on this important topic with applications to transition-metal complexes has appeared in the literature [12–14]. Another restriction of the scope of this paper is the focus on isotropic shielding and spin–spin coupling constants. The tensorial properties of nuclear shielding and spin–spin coupling are only very briefly mentioned and the reader is referred to Refs. [15, 16] for further details.

In the methodology part, atomic units with $e = 1$,$m_e = 1$,$h = 2\pi$,$4\pi\varepsilon_0 = 1$, and the speed of light $c = 137.03599976(50)$ are employed. Thus prefactors containing e,m_e,$4\pi\varepsilon_0$ and $\hbar = h/(2\pi)$ are often dropped, whereas the speed of light, c, is explicitly included. Equations referring to magnetic properties have been converted from SI to atomic units. The fine structure constant α is in atomic units equal to c^{-1}. The latter occurs in the equations indicating the relative magnitude of the hyperfine terms in the Hamiltonian as compared with relativistic corrections of $\mathcal{O}(c^{-2})$ and the nonrelativistic terms $\mathcal{O}(c^0)$.

2 Methodological Aspects of NMR Computations

2.1 Nuclear Magnetic Resonance

The presence of magnetic moments $\boldsymbol{\mu}_A$,$\boldsymbol{\mu}_B, \ldots$ of nuclei A,$B, \ldots$ in a molecule are responsible for the two observables of the NMR experiment that are most frequently utilized in chemical applications. They are physically observed in form of quantized energy differences ΔE that can be measured very precisely. These two observables are the nuclear shielding tensor $\boldsymbol{\sigma}_A$ for nucleus A and the so-called indirect reduced coupling tensor K_{AB} for a pair of nuclei A,B. Both $\boldsymbol{\sigma}_A$ and K_{AB} are second-rank tensors that are defined via the phenomenological Hamiltonians

$$H^{\text{spin}} = -\boldsymbol{\mu}_A(1 - \boldsymbol{\sigma}_A)\boldsymbol{B}^{\text{ext}} \tag{1a}$$

for the shielding tensor and

$$H^{\text{spin}} = \boldsymbol{\mu}_A \mathsf{K}_{AB} \boldsymbol{\mu}_B \tag{1b}$$

for the spin–spin coupling. $\boldsymbol{B}^{\text{ext}}$ is the external magnetic field. Equations (1a) and (1b) do not contain any electronic degrees of freedom. The external field as well as the magnetic moments of the nuclei are fixed, known quantities which are experimentally determined, not quantum operators that act on electronic variables. $\boldsymbol{\mu}_A$ is related to the spin $\boldsymbol{I}_A$ of nucleus A by

$$\boldsymbol{\mu}_A = \hbar\gamma_A \boldsymbol{I}_A \quad , \tag{2}$$

with γ_A being the magneto-gyric ratio of A. The γ's enter the relation between the reduced coupling tensor (in SI units of tesla squared per joule) and the experimentally determined J tensor (in units of hertz):

$$\mathsf{J}_{AB} = \frac{\hbar}{2\pi}\gamma_A\gamma_B\mathsf{K}_{AB} \quad . \tag{3}$$

The rotational averages of the shielding and coupling tensors are the most important experimentally determined parameters. They are given by one third of the trace (sum of diagonal elements) of the respective tensors and will be denoted with σ_A or $\sigma(A)$ (isotropic shielding constant) and K_{AB} or $K(A\text{-}B)$ (reduced spin–spin coupling constant, similar for J) in the following.

There is also a direct coupling tensor D_{AB} for a pair of nuclei which is independent of the electronic structure of the molecule under consideration. Its elements depend on the internuclear distances and the orientation of the molecule with respect to the chosen coordinate system and can thus be employed for structural information [15]. It also becomes important in solid-state or liquid-crystal NMR experiments, or in cases when the molecules are (partially) aligned by a strong external field. However, for rapidly and freely rotating molecules (gas phase or solution) D_{AB} does not contribute to the observed coupling and shielding constants and shall therefore not be considered further. The indirect coupling constant as well as the nuclear shielding arise from the presence of the electrons in the molecule and thus carry a wealth of chemical information which is now routinely being extracted from convenient measurements in solution.

2.2 NMR Parameters Defined as Second-Order Energy Perturbations

In this and the following subsections we assume either knowledge of a hypothetical exact wavefunction or electron density, or when referring to existing methodology, knowledge of an approximate wavefunction or electron density that has been determined such as to minimize the energy (variational approach). This is not a necessary condition in order to calculate molecular properties in general and NMR parameters in particular, however it facilitates qualitative discussions and interpretations based on wavefunction or electron density perturbations. For a more general account of calculations of energy derivatives the reader is referred to Ref. [17].

In first-principles quantum theories of molecules, a system is characterized by its potential (nuclear coordinates and charges), its number of electrons and its kinematics (the form of the operators for the kinetic energy and the electronic repulsion, for instance nonrelativistic versus relativistic expressions). Therefore, a transition-metal complex is formally treated the same way as other molecules. The shielding and indirect spin-spin coupling tensors are thereby obtained as double first-order perturbations of the total energy E of the molecule. In variational wavefunction based methods,

$$E = \langle \Psi | \hat{H} | \Psi \rangle \quad , \tag{4}$$

with Ψ being the normalized wavefunction. In Kohn–Sham DFT,

$$E = E[\rho] \quad ; \quad \rho = \sum_i^{\mathrm{occ}} \varphi_i^* \varphi_i \quad , \tag{5}$$

with ρ being the electron density and the φ_i the occupied Kohn–Sham molecular orbitals (MOs). $E[\rho]$ indicates that E is a functional of the density. It should be noted that in Kohn–Sham DFT the kinetic energy is not calculated from the density. For NMR calculations, the two perturbation parameters are $\boldsymbol{\mu}_A$, and $\boldsymbol{\mu}_B$ or $\boldsymbol{B}^{\mathrm{ext}}$, respectively, i.e.

$$\boldsymbol{\sigma}_A = \left. \frac{\partial^2 E}{\partial \boldsymbol{\mu}_A \partial \boldsymbol{B}^{\mathrm{ext}}} \right|_{\boldsymbol{\mu}_A=0, \boldsymbol{B}^{\mathrm{ext}}=0} \tag{6a}$$

and

$$\mathsf{K}_{AB} = \left. \frac{\partial^2 E}{\partial \boldsymbol{\mu}_A \partial \boldsymbol{\mu}_B} \right|_{\boldsymbol{\mu}_A=0, \boldsymbol{\mu}_B=0} \quad . \tag{6b}$$

In Eq. (6a) the energy is supposed not to contain the classical nuclear Zeeman term $-\boldsymbol{\mu}_A \boldsymbol{B}^{\mathrm{ext}}$ which accounts for the 1 in Eq. (1a). Note the formal agreement of Eqs. (6a) and (6b) with the respective derivatives of Eqs. (1a) and (1b). These two equations are obtained first by deriving an expression of the molecular energy E which includes all terms in the molecular quantum Hamiltonian $\hat{H}$ that are related to the nuclear magnetic moments and the external field $\boldsymbol{B}^{\mathrm{ext}}$, and second by considering their interaction with the magnetic moments arising from the electronic spins and their motion through space. Accordingly, Ψ or ρ must be understood as being dependent on $\boldsymbol{\mu}_A$, $\boldsymbol{\mu}_B$ and $\boldsymbol{B}^{\mathrm{ext}}$. Integration over all electronic degrees of freedom yields expressions of the same formal structure as Eqs. (1a) and (1b). Thus the combined terms of explicit expressions for Eq. (4) or Eq. (5) that are bilinear in $\boldsymbol{\mu}_A$ and $\boldsymbol{\mu}_B$ correspond to Eq. (1b) and yield K_{AB} upon differentiation, whereas the terms that are bilinear in $\boldsymbol{\mu}_A$ and $\boldsymbol{B}^{\mathrm{ext}}$ correspond to Eq. (1a) and yield $\boldsymbol{\sigma}_A$.

Taking the derivatives of E with respect to $\boldsymbol{\mu}_A$ and $\boldsymbol{\mu}_B$ or $\boldsymbol{B}^{\mathrm{ext}}$ involves first taking the derivatives of the operators that constitute the Hamiltonian $\hat{H}$ or the Kohn–Sham density functional and that are dependent on $\boldsymbol{\mu}_A$ and/or $\boldsymbol{\mu}_B$

and/or $\boldsymbol{B}^{\text{ext}}$. They can be derived once a particular (relativistic or nonrelativistic) Hamiltonian or density functional has been adopted. Second, the presence of the magnetic perturbations in the Hamiltonian causes (small) changes in the electronic structure (i.e. Ψ or ρ) that need to be taken into account to first order in order to calculate the second-order perturbations of the energy. Thus, derivatives of the wavefunction Ψ or the electron density ρ , with respect to at least one of the perturbations, are additionally required. Here, the particular form of energy expression within different ab-initio methods, but also DFT (e.g., density versus current-density functionals), is crucial for obtaining the correct equations for these perturbations.

The notation in Eqs. (6a) and (6b) corresponds to the usual notation of perturbation theory [18] in which it is understood that the derivatives of the Hamiltonian as well as all variational and nonvariational parameters (i.e., including the dependence of the wavefunction on the perturbations) are taken. Some authors prefer to write Eqs. (6a) and (6b) as total derivatives of E in order to indicate that all dependencies of E on the perturbation parameters must be considered [19, 20]. Ultimately one needs to consider an explicit energy expression in which no hidden dependencies on any of the perturbations are left. This also includes a possible dependence of the basis set on the perturbation.

2.3 The Gauge Problem

Any DFT computations of NMR properties of molecules with present-day program codes requires either the use of a finite basis set or the use of a finite number of points on a grid, or both, to represent the MOs. Obviously, no complete (infinite size) basis can be employed, which has some important consequences for basis set calculations of σ_A, generally known as the "gauge dependence" or "gauge origin" problem. This applies to other molecular properties related to a $\boldsymbol{B}^{\text{ext}}$ perturbation also. It is introduced via the magnetic vector potential $\boldsymbol{A}^{\text{ext}}$ related to $\boldsymbol{B}^{\text{ext}} = \boldsymbol{\nabla} \times \boldsymbol{A}^{\text{ext}}$ which directly enters the magnetic terms in the Hamiltonian. Here we do not want to elaborate on the origin of the gauge problem but refer to some recent overview articles for details [19, 21]. See also McWeeny's textbook [22]. For the purpose of this article it is sufficient to mention that a gauge dependence of σ_A arises from using an incomplete basis set. It makes the shielding tensor more or less strongly dependent on the chosen origin of the coordinate system for the computation. This renders a comparison of nuclear shieldings within a molecule, or between nuclei of different molecules that are not put at the same coordinates, questionable. Generally, the better the basis set is, the less pronounced is the origin dependence of σ_A due to the gauge problem; however, the convergence of the results with respect to the basis size is slow and also small differences of chemical shifts are meaningful, important and experimentally detectable.

Additional terms in explicit expressions for the shielding tensor, either in the atomic basis set or at some intermediate level, are deliberately introduced in order to remove or reduce the origin dependence of σ_A. The larger the basis set becomes, the less important become these contributions. Curing the gauge problem in the formalism has the additional advantage that this effectively leads to well-converged results for shielding tensors even with moderately sized basis sets [17], though a program implementation is somewhat more involved. There are a number of different methodologies available that are concerned with the origin dependence, among them the use of gauge including atomic orbitals (GIAO), of an individual gauge of localized orbitals (IGLO), a localized orbital / local origin (LORG), or applying a continuous set of gauge transformations (CSGT). See Sect. 3 for references. It is now widely recognized that nuclear shielding computations should in some way include gauge correction terms, with the GIAO method being among the most effective and widespread. In this method, the atomic basis functions are chosen to depend on $\boldsymbol{B}^{\text{ext}}$ in such a way that σ_A becomes origin-independent even when using a minimal basis. [1] Other methods solve the gauge problem in a different way, but they all have in common that they lead to additional contributions in σ_A which depend on the coordinates of the nucleus and remove or reduce the origin-dependency. Such approaches are commonly referred to as distributed gauge origin methods, as compared with adopting a common gauge origin.

2.4 The (2n+1) Theorem and the Interchange Theorem

When calculating the second derivatives of E in a variational formalism, use can be made of the so-called $(2n + 1)$ theorem and the interchange theorem [18]. The $(2n + 1)$ theorem states that the availability of the nth order perturbation of the wavefunction or electron density allows the calculation of the perturbed energy up to order $2n + 1$. The interchange theorem states that the order in which the derivatives are taken in Eqs. (6a) and (6b) is arbitrary. This means that only the derivative of Ψ or ρ with respect to μ_A or $\boldsymbol{B}^{\text{ext}}$ is needed, not both, in order to calculate the nuclear shielding. The respective perturbed Ψ or ρ is in turn calculated from expanding their defining equation(s) (Kohn–Sham equations, Hartree–Fock equations, ..., Schrödinger equation) in a power series with respect to the perturbation parameters and collecting terms linear in the chosen perturbation. The solution of this equation or these equations represents the main computational task in the determination of σ_A or K_{AB} and is often

[1] It is a different question whether results obtained with small basis sets, even if origin-independent, are meaningful at all

similarly expensive as finding the unperturbed (zeroth-order) wavefunction or electron density. The latter must be known prior to carrying out a NMR calculation. For reasons of computational efficiency, in nuclear shielding computations the perturbation equations or equation is solved with $\boldsymbol{B}^{\mathrm{ext}}$ as the perturbation, since the shielding tensors for all the nuclei in the molecule can subsequently be determined without too much additional computational cost. In calculations of K_{AB}, a perturbed Ψ or ρ must be available for one of the nuclei for each pair A,B of interest. The result is independent of which nucleus is the "perturbing" one, i.e., for which nucleus the perturbed Ψ or ρ is calculated.

2.5 Perturbation Operators

Let $\hat{H}^{(\boldsymbol{\mu}_A)}$ collectively denote the all "magnetic" operators that contribute to the first derivative of $\hat{H}$ with respect to $\boldsymbol{\mu}_A$ at $\boldsymbol{\mu}_A = \boldsymbol{\mu}_B = \boldsymbol{B}^{\mathrm{ext}} = 0$. Accordingly, $\hat{H}^{(\boldsymbol{\mu}_B)}$ and $\hat{H}^{(\boldsymbol{B}^{\mathrm{ext}})}$ are the derivatives with respect to $\boldsymbol{\mu}_B$ and $\boldsymbol{B}^{\mathrm{ext}}$, respectively, and $\hat{H}^{(\boldsymbol{\mu}_A,\boldsymbol{\mu}_B)}$ etc. denote the mixed second derivatives. Further, $\hat{h}^{(\boldsymbol{\mu}_A)}$ shall denote a respective one-electron operator, i.e., $\hat{H}^{(\boldsymbol{\mu}_A)} = \sum_{i=1}^{N} \hat{h}^{(\boldsymbol{\mu}_A)}(i)$ + two-electron terms, if applicable. N is the number of electrons in the system. A number of nonrelativistic one-electron terms in the molecular Hamiltonian that yield these perturbation operators upon differentiation are listed in in Table 1 and these in turn enter the expressions for $\boldsymbol{\sigma}_A$ and K_{AB}. Relativistic operators are discussed in Refs. [6–8]. There, the reader can also find literature references where the perturbation operators for specific relativistic methodologies are explicitly listed.

2.6 The Sum-Over-States Formula

Let us for reasons of simplicity not consider any particular expression for the energy E at this point. As already mentioned, for the computation of the nuclear shielding $\boldsymbol{\sigma}_A$, one needs $\Psi^{(\boldsymbol{B}^{\mathrm{ext}})}$, the first-order derivative of the wavefunction with respect to the external field at $\boldsymbol{\mu}_A = 0$,$\boldsymbol{B}^{\mathrm{ext}} = 0$. For the computation of the coupling constant, knowledge of $\Psi^{(\boldsymbol{\mu}_A)}$ is required. Let Ψ_0 be the unperturbed wavefunction of the ground state of the molecule, or the reference state, for which the NMR property is to be calculated. The shielding tensor is then given by differentiating Eq. (4) in the sense of Eq. (6a), which directly yields

$$\begin{aligned}\boldsymbol{\sigma}_A &= \langle \Psi_0 | \hat{H}^{(\boldsymbol{\mu}_A,\boldsymbol{B}^{\mathrm{ext}})} | \Psi_0 \rangle + \langle \Psi^{(\boldsymbol{B}^{\mathrm{ext}})} | \hat{H}^{(\boldsymbol{\mu}_A)} | \Psi_0 \rangle + \langle \Psi_0 | \hat{H}^{(\boldsymbol{\mu}_A)} | \Psi^{(\boldsymbol{B}^{\mathrm{ext}})} \rangle \\ &= \langle \Psi_0 | \hat{H}^{(\boldsymbol{\mu}_A,\boldsymbol{B}^{\mathrm{ext}})} | \Psi_0 \rangle + 2\,\mathrm{Re} \langle \Psi^{(\boldsymbol{B}^{\mathrm{ext}})} | \hat{H}^{(\boldsymbol{\mu}_A)} | \Psi_0 \rangle \quad .\end{aligned} \tag{7a}$$

Table 1 Nonrelativistic one-electron "magnetic" terms in the Hamiltonian. Their derivatives with respect to $\boldsymbol{\mu}_A$ and/or $\boldsymbol{B}^{\text{ext}}$ or $\boldsymbol{\mu}_B$ enter the expressions for the nuclear shielding and spin-spin coupling tensors via the perturbation operators $\hat{h}^{(\boldsymbol{\mu}_A)}$,$\hat{h}^{(\boldsymbol{\mu}_B)}$,$\hat{h}^{(\boldsymbol{B}^{\text{ext}})}$,$\hat{h}^{(\boldsymbol{\mu}_A,\boldsymbol{B}^{\text{ext}})}$ and $\hat{h}^{(\boldsymbol{\mu}_A,\boldsymbol{\mu}_B)}$. $\hat{\boldsymbol{S}}$ is the spin-operator for an electron, $\boldsymbol{r}_A$ a distance vector with respect to nucleus A etc.

Operator	Name	Contributes to
$\hat{h}^{OZ} = -\frac{i}{2}\boldsymbol{B}^{\text{ext}} \cdot (\boldsymbol{r} \times \boldsymbol{\nabla})$	Orbital Zeeman	$\hat{h}^{(\boldsymbol{B}^{\text{ext}})}$
$\hat{h}^{SZ} = \boldsymbol{B}^{\text{ext}} \cdot \hat{\boldsymbol{S}}$	Spin-Zeeman [a]	$\hat{h}^{(\boldsymbol{B}^{\text{ext}})}$
$\hat{h}^{OP} = -\frac{i}{c^2}\sum_A \boldsymbol{\mu}_A(\frac{\boldsymbol{r}_A}{r_A^3} \times \boldsymbol{\nabla})$	Paramagnetic [b] Orbital	$\hat{h}^{(\boldsymbol{\mu}_A)}$,$\hat{h}^{(\boldsymbol{\mu}_B)}$
$\hat{h}^{FC} + \hat{h}^{SD} = \frac{1}{c^2}\sum_A \hat{\boldsymbol{S}}[\boldsymbol{\mu}_A(\boldsymbol{\nabla} \cdot \frac{\boldsymbol{r}_A}{r_A^3}) -(\boldsymbol{\mu}_A \cdot \boldsymbol{\nabla})\frac{\boldsymbol{r}_A}{r_A^3}]$		$\hat{h}^{(\boldsymbol{\mu}_A)}$,$\hat{h}^{(\boldsymbol{\mu}_B)}$
$\hat{h}^{FC} = \frac{8\pi}{3c^2}\sum_A \delta(\boldsymbol{r}_A)\boldsymbol{\mu}_A \cdot \hat{\boldsymbol{S}}$	Fermi Contact [a]	$\hat{h}^{(\boldsymbol{\mu}_A)}$,$\hat{h}^{(\boldsymbol{\mu}_B)}$
$\hat{h}^{SD} = \frac{1}{c^2}\sum_A r_A^{-5}\left[3(\hat{\boldsymbol{S}} \cdot \boldsymbol{r}_A)(\boldsymbol{\mu}_A \cdot \boldsymbol{r}_A) -r_A^2\boldsymbol{\mu}_A \cdot \hat{\boldsymbol{S}}\right]$	Spin-Dipole [a]	$\hat{h}^{(\boldsymbol{\mu}_A)}$,$\hat{h}^{(\boldsymbol{\mu}_B)}$
$\hat{h}^{DS} = \frac{1}{2c^2}\sum_A [(\boldsymbol{\mu}_A \cdot \boldsymbol{B}^{\text{ext}})(\frac{\boldsymbol{r}_A}{r_A^3} \cdot \boldsymbol{r}) -(\boldsymbol{\mu}_A \cdot \boldsymbol{r})(\boldsymbol{B}^{\text{ext}} \cdot \frac{\boldsymbol{r}_A}{r_A^3})]$	Diamagnetic Shielding	$\hat{h}^{(\boldsymbol{\mu}_A,\boldsymbol{B}^{\text{ext}})}$
$\hat{h}^{OD} = \frac{1}{2c^4}\sum_{B \neq A} r_A^{-3}r_B^{-3}[(\boldsymbol{\mu}_A \cdot \boldsymbol{\mu}_B)(\boldsymbol{r}_A \cdot \boldsymbol{r}_B) -(\boldsymbol{\mu}_A \cdot \boldsymbol{r}_B)(\boldsymbol{\mu}_B \cdot \boldsymbol{r}_A)]$	Diamagnetic [c] Orbital	$\hat{h}^{(\boldsymbol{\mu}_A,\boldsymbol{\mu}_B)}$

[a] No contribution to $\boldsymbol{\sigma}_A$ in nonrelativistic or scalar relativistic calculations excluding spin-orbit coupling
[b] Also called "Paramagnetic Spin-Orbit" (PSO), referring to nuclear spins and electronic orbits
[c] Also called "Diamagnetic Spin-Orbit" (DSO), see also [b]

Here, the terminology of Sect. 2.5 has been adopted. The first term in Eq. (7a) is called the diamagnetic shielding; the second term is called the paramagnetic shielding. This subdivision depends on the choice of gauge of the magnetic vector potential (here Coulomb gauge) and refers to the nonrelativistic theory of nuclear shielding. Additional contributions might be contained in the expression if $\hat{H}$ and E refer to a relativistic formalism, in particular those due to the electronic spin-orbit coupling. The latter belong to the "paramagnetic" type. In a four-component formalism there is no explicit diamagnetic term in the respective equation.

The reduced indirect spin–spin coupling tensor is given by

$$\begin{aligned}\mathsf{K}_{AB} &= \langle\Psi_0|\hat{H}^{(\mu_A,\mu_B)}|\Psi_0\rangle + \langle\Psi^{(\mu_A)}|\hat{H}^{(\mu_B)}|\Psi^{(0)}\rangle + \langle\Psi_0|\hat{H}^{(\mu_B)}|\Psi^{(\mu_A)}\rangle \\ &= \langle\Psi_0|\hat{H}^{(\mu_A,\mu_B)}|\Psi_0\rangle + 2\,\mathrm{Re}\langle\Psi^{(\mu_A)}|\hat{H}^{(\mu_B)}|\Psi_0\rangle \quad .\end{aligned} \tag{7b}$$

Again, the first term in Eq. (7b) is commonly referred to as the diamagnetic coupling contribution, diamagnetic orbital (OD) term, or diamagnetic (nuclear) spin - (electronic) orbit term. The second term contains, in a nonrelativistic theory, the Fermi contact (FC) term, the spin–dipole (SD) and the FC–SD cross term, as well as the paramagnetic orbital (OP) term. In a relativistic theory, spin–orbit contributions are further included (e.g. the FC–OP cross term). In deriving both expressions, Eqs. (7a) and (7b), use has been made of the Hellmann–Feynman theorem [23] which allows us to avoid the computation of a second derivative of the wavefunction, as mentioned before in the context of the $(2n+1)$ theorem. Note the equal formal structure of the two equations, which is in fact the same for all time-independent second-order perturbations of the energy [4].

The operators as well as the unperturbed wavefunction are supposed to be known, thus the only unknown quantities in Eqs. (7a) and (7b) are $\Psi^{(\boldsymbol{B}^{\mathrm{ext}})}$ and $\Psi^{(\mu_A)}$, respectively. In the sum-over-states (SOS) approach, all the unperturbed excited-state wavefunctions Ψ_j are now used as a convenient complete orthonormal basis set in which to expand the perturbed wavefunctions of the ground state. By considering only first-order terms in the (relativistic or nonrelativistic) wave equation, the following expression for $\Psi^{(\boldsymbol{B}^{\mathrm{ext}})}$ is readily derived:

$$\Psi^{(\boldsymbol{B}^{\mathrm{ext}})} = \sum_{j\neq 0} \frac{\langle\Psi_j|\hat{H}^{(\boldsymbol{B}^{\mathrm{ext}})}|\Psi_0\rangle}{E_0 - E_j}\Psi_j \tag{8}$$

The expression for $\Psi^{(\mu_A)}$ is obtained by changing $\hat{H}^{(\boldsymbol{B}^{\mathrm{ext}})}$ to $\hat{H}^{(\mu_A)}$. When these results are substituted in Eqs. (7a) and (7b), the following (in principle exact) SOS expressions for the NMR observables are obtained:

$$\sigma_A = \langle\Psi_0|\hat{H}^{(\mu_A,\boldsymbol{B}^{\mathrm{ext}})}|\Psi_0\rangle + 2\sum_{j\neq 0}\mathrm{Re}\,\frac{\langle\Psi_0|\hat{H}^{(\boldsymbol{B}^{\mathrm{ext}})}|\Psi_j\rangle\langle\Psi_j|\hat{H}^{(\mu_B)}|\Psi_0\rangle}{E_0 - E_j} \tag{9a}$$

$$\mathsf{K}_{AB} = \langle\Psi_0|\hat{H}^{(\mu_A,\mu_B)}|\Psi_0\rangle + 2\sum_{j\neq 0}\mathrm{Re}\,\frac{\langle\Psi_0|\hat{H}^{(\mu_A)}|\Psi_j\rangle\langle\Psi_j|\hat{H}^{(\mu_B)}|\Psi_0\rangle}{E_0 - E_j} \tag{9b}$$

Of course in practice not all, usually not even one, of the excited-state wavefunctions are known and therefore, Eqs. (9a) and (9b) as they stand do not have much practical importance; however they greatly facilitate interpretations. For instance, systems with low-lying exited states are likely to have larger values for their NMR observables (and other second-order properties). Trends among related compounds for magnetic properties can often be rationalized by trends observed for the excitation energies. For instance, ^{95}Mo chemical shifts in a number of complexes were shown in Ref. [24] to correlate very well with the energy of the lowest-lying magnetically allowed excitation. By knowing the form of the perturbation operators it is also often possible to make qualitative predictions on the size of the matrix elements in the numerator of the SOS expression, which might yield useful interpretations. On the basis of nonrelativistic formalism it can easily be deduced from symmetry arguments that in atoms there is no paramagnetic contribution to the shielding tensor, and that in linear molecules the paramagnetic principal components of the $\boldsymbol{\sigma}_A$ and K_{AB} tensors are zero along the molecular axis.

2.7 Density Functional Theory

In Kohn–Sham DFT based approaches, expressions that are of similar structure as Eqs. (9a) and (9b) are obtained, but in the form of contributions from all occupied Kohn–Sham MOs φ_i. The excited-state wavefunctions are at the same time formally replaced by the unoccupied MOs, and the many-electron perturbation operators $\hat{H}^{(\mu_A)}$, etc. by their one-electron counterparts $\hat{h}^{(\mu_A)}$, etc. Orbital energies ε_i and ε_a formally substitute the total energies of the states (see later). Thus, similar interpretations of NMR parameters can be worked out in which the highest occupied MO–lowest unoccupied MO gap (HLG) plays a highly important role. It must be emphasized, though, that there is no one-to-one correspondence between the excited states of the SOS equations and the unoccupied orbitals which enter the DFT expressions, nor between excitation energies and orbital energy differences, i.e., there are no one-determinantal "wavefunctions" in Kohn–Sham DFT perturbation theory which approximate the reference and excited states.

In formal analogy to the perturbed $\Psi^{(\mu_A)}$, etc., in Kohn–Sham DFT there are the perturbations $\varphi_i^{(\mu_A)}$, etc. of the occupied MOs that are in turn represented in the complete orthonormal basis of all unperturbed MOs. The occupied orbitals enter the expressions for the effective Hartree plus exchange-correlation (HXC) potential V_{HXC} of the molecule which originates from the electron–electron Coulomb interaction and the Pauli principle. Because of the orbitals being perturbed, in addition to the "external" perturbation (magnetic field or nuclear spin) there is also a "coupled" perturbation contribution $V_{\mathrm{HXC}}^{(\boldsymbol{B}^{\mathrm{ext}})}$ or $V_{\mathrm{HXC}}^{(\mu_A)}$ (similar to coupled Hartree–Fock theory). Considering such terms, the DFT equations for $\boldsymbol{\sigma}_A$ and K_{AB} read

$$\begin{aligned}\sigma_A &= \sum_i^{\text{occ}} \langle \varphi_i | \hat{h}^{(\mu_A, B^{\text{ext}})} | \varphi_i \rangle \\ &+2 \sum_i^{\text{occ}} \sum_a^{\text{unocc}} \text{Re}\, \frac{\langle \varphi_i | \hat{h}^{(B^{\text{ext}})} + V_{HXC}^{(B^{\text{ext}})} | \varphi_a \rangle \langle \varphi_a | \hat{h}^{(\mu_B)} | \varphi_i \rangle}{\varepsilon_i - \varepsilon_a}\end{aligned} \tag{10a}$$

$$\begin{aligned}\mathsf{K}_{AB} &= \sum_i^{\text{occ}} \langle \varphi_i | \hat{h}^{(\mu_A, \mu_B)} | \varphi_i \rangle \\ &+2 \sum_i^{\text{occ}} \sum_a^{\text{unocc}} \text{Re}\, \frac{\langle \varphi_i | \hat{h}^{(\mu_A)} + V_{HXC}^{(\mu_A)} | \varphi_a \rangle \langle \varphi_a | \hat{h}^{(\mu_B)} | \varphi_i \rangle}{\varepsilon_i - \varepsilon_a}\end{aligned} \tag{10b}$$

and are in principle exact, though as usual in DFT the XC contribution is only known to some approximation. An exact (complete basis) representation of the orbitals is assumed, therefore Eqs. (10a) and (10b) do not include any of the aforementioned "gauge terms". With functionals that do not depend on the current-density the contribution $V_{\text{HXC}}^{(B^{\text{ext}})}$ is zero. We will in the following refer to such cases as "uncoupled" DFT perturbation methods. It should be noted that this notion refers to nonhybrid functionals that depend on the density, its gradients and higher derivatives, or the kinetic energy density, but not to hybrid functionals that include Hartree–Fock exchange. In the latter case a coupled perturbed $V_{\text{HXC}}^{(B^{\text{ext}})}$ corresponding to the fraction of Hartree–Fock exchange in the hybrid functional is usually included in the NMR computations. Regarding fully coupled perturbed current-density functional methods applied to NMR calculations not much data are available in the literature. It appears that the contributions from the current-density dependent part of the functional are negligible [25], which somewhat justifies the common uncoupled approach. On the other hand, so far only comparatively simple current-density functionals have been applied in such investigations. It is presently not clear whether the current-density-dependent coupled terms will gain importance when more sophisticated functionals are applied or whether they can become large, for instance, for other compounds that were tested so far (e.g., they might be important for chemical shifts of transition-metals). So-called SOS DFT has been devised [26, 27] in which the orbital energy denominators of the DFT expression are replaced by approximate excitation energies, in order to cure some of the deficiencies of the various available nonhybrid XC potentials. This approach has found some justification by producing more accurate results in a number of applications, but has also been criticized for its semiempirical nature. Recently, similar expressions have been obtained not by resorting to the SOS argument but from an argument based on time-dependent DFT methods, in an attempt to derive equivalent time-dependent DFT analogs for the SOS expression for electric and magnetic perturbations [28,29]. Attempts have further been made to obtain improved NMR parameters by self-interaction-corrected DFT [30, 31] and other "nonstandard" Kohn–Sham potentials [32]. There are currently not enough data available in order to judge their performance for transition-metal complexes.

As already mentioned, the MOs in Eqs. (10a) and (10b) are assumed to be exact, i.e., represented at the complete basis set limit. In practice, the Kohn–Sham equations are converted into the respective self-consistent-field matrix equations in the basis, and the perturbation treatment is carried out from there. Any dependence of the basis set (GIAOs) on the perturbation ($\boldsymbol{B}^{\text{ext}}$ in the case of shielding tensors) is in this way naturally covered.

2.8 Singlet and Triplet Perturbations, Spin–Orbit Coupling

Some of the perturbation operators in Table 1 do not depend on the electron spin. As a consequence, integration over spin can be carried out in Eqs. (9a) and (9b) for all terms that contain only these spin-free operators. They yield zero in a nonrelativistic or scalar relativistic framework (i.e., without spin–orbit coupling) unless Ψ_j is a singlet excited state (Ψ_0 is supposed to be a singlet closed-shell ground state). Similar arguments apply to the contributions from spin-up and spin-down MOs in Eqs. (10a) and (10b). Such perturbations are often referred to as "singlet perturbations". The spin-dependent perturbations are referred to as "triplet perturbations" and only yield nonzero contributions from excited triplet states in the perturbed wavefunction, see Eq. (8). This means that such a perturbation also causes a spin-polarization of the closed-shell ground state. The FC operator in Table 1 is an example. Without spin–orbit coupling, for a closed shell molecule all cross terms where in the SOS equation or in Eqs. (10a) and (10b) one of the perturbation operators is a singlet and the other one is a triplet perturbation exactly cancel owing to contributions of equal magnitude but opposite sign from the spin-up and spin-down orbitals, respectively. Therefore, the FC and SD operators do not contribute to the shielding tensor at this level of theory.

If spin–orbit coupling is present, the orbitals or the density of the system can in general not be divided anymore into purely spin-up or spin-down contributions but there are contributions from both. The action of a singlet perturbation, for example, $\hat{H}^{(\boldsymbol{B}^{\text{ext}})}$, then leads to different results for the spin-up and the spin-down portions of the orbitals. This in turn causes a spin-density in the molecule to be induced by a singlet perturbation. For instance, when spin–orbit coupling is considered the external field (orbital Zeeman operator) yields cross terms with the FC and SD terms, and represents the most important spin–orbit effect on the nuclear shielding [33–35]. Likewise, for the spin–spin coupling tensor there occur cross terms between FC or SD and the singlet OP term.

2.9 Relativistic Methods

So far no specification has been made whether the energy expression on which the NMR calculation is based is relativistic or not. It has long been known that electrons "move very fast" and, in molecules with heavy nuclei, reach a substantial portion of the speed of light. Thus, the effects of Einstein's special relativity must be considered in their theoretical description. The reader is referred to the literature for accounts of relativistic methodology for molecules [36–40] and the computation of molecular properties within such formalisms [4]. For review articles that are specifically devoted to relativistic NMR computations, see Refs. [6–8]. It is important to know that relativistic effects are stronger for systems with heavy nuclei (the leading order effects increase approximately as Z^2). It depends on the accuracy of the calculation — and which quantity is calculated — whether a relativistic calculation must be carried out or not. For the 5*d* transition-metals, relativistic corrections should generally be considered, independent of the goal of the calculation (provided the goal is not to produce qualitatively wrong answers).

For the present discussion, it is of importance that a relativistic NMR computation is carried out if E in Eq. (4) is defined and calculated within a relativistic framework, otherwise it is nonrelativistic. Another approach is to treat relativity as an additional perturbation and to evaluate relativistic corrections to $\boldsymbol{\sigma}_A$ and K_{AB} individually in the form of derivatives of Eqs. (6a) and (6b) with respect to a suitably chosen relativistic perturbation parameter (usually c^{-2}), i.e. first a nonrelativistic NMR observable is calculated and then correction terms are obtained in the form of

$$\Delta^{\text{rel}}\boldsymbol{\sigma}_A = c^{-2} \cdot \left. \frac{\partial^3 E}{\partial \boldsymbol{B}^{\text{ext}} \partial \boldsymbol{\mu}_A \partial c^{-2}} \right|_{c\to\infty, \boldsymbol{\mu}_A=0, \boldsymbol{B}^{\text{ext}}=0} + \mathcal{O}(c^{-4}) \tag{11}$$

and similarly for K_{AB}. In the variational relativistic approaches, the second derivatives Eqs. (6a) and (6b) of E yield unseparated the nonrelativistic quantities plus all relativistic corrections that the scheme covers. Perturbational approaches can be advantageous for interpretational purposes [clear separation of leading (c^{-2}) and higher order (c^{-4},c^{-6} ...) relativistic corrections] but it should be kept in mind that for the heaviest NMR nuclei the relativistic effects on $\boldsymbol{\sigma}$ can be huge and will not be fully covered by the terms linear in c^{-2}. This gives variational relativistic methods the advantage of relative conceptual simplicity over higher-order relativistic perturbational approaches when computing $\boldsymbol{\sigma}_A$ or K_{AB} for a heavy nucleus or its neighboring atoms. If desired, a nonrelativistic calculation can always be carried out in addition in order to estimate the magnitude of the relativistic contributions.

The computation of NMR observables within a relativistic formalism has some important consequences. The most obvious is that the Hamiltonian of the system — and thus the perturbation operators — are different in a variational relativistic scheme (i.e., if relativity is not treated as a perturbation. Otherwise, the nonrelativistic operators of Table 1 occur in the leading-order terms). There are a number of different methodologies available for carrying out relativistic computations in a variationally stable manner. Each of them requires the implementation of a specific set of perturbation operators for the purpose of NMR calculations. It is very important to keep the method internally consistent in order to avoid singularities, variational collapse, and meaningless results. For instance, for point nuclei the FC operator yields arbitrarily large matrix elements in conjunction with relativistic MOs, whereas its relativistic analogs stay finite [8].

To the author's knowledge, four-component relativistic computations have so far not been carried out for NMR parameters of transition-metal complexes, which is likely due to the formidable computational effort since respective program codes exist. Two-component methods can be more efficient and have been applied to a large range of transition-metal complexes. In this paper, two-component methods are classified to include the important effects of spin-orbit coupling, some of which will be discussed in Sect. 4. If spin-orbit coupling is neglected in a two-component Hamiltonian, the approach turns into a one-component, so-called scalar relativistic one. Presently, for transition-metal systems the most frequently applied two-component relativistic DFT NMR method is based on the zeroth-order regular approximation (ZORA) [41–46]. Nuclear shieldings have further successfully been computed based on the second-order Douglas–Kroll–Hess (DKH2) transformed Hamiltonian [40, 47–50]. A variationally not stable approach using the two-component Pauli Hamiltonian in conjunction with frozen cores has for some years been utilized [2, 51–53], but is now depreciated in favor of variationally stable methods. In the following, this method will be designated by vPSC if only scalar relativistic effects are considered, otherwise by vPSCSO. Perturbational results based on Eq. (11) will be mentioned in Sect. 4 as well. They have so far employed the Pauli operator and will be designated by pPSC or pPSCSO, respectively.

Besides these direct relativistic methods, the relativistic effects on a heavy atom's valence shell can also be incorporated in an otherwise nonrelativistic computation by employing effective core potentials (ECPs) for the heavy atom. As long as the core shell of the heavy elements is not of particular interest (i.e., for structure, energetics and most properties except electric and magnetic properties related to the heavy nucleus or the core shells of the heavy atom) such computations are very efficient and accurate [54]. For example, ligand chemical shift or coupling constants between the ligands of a heavy metal can be calculated in this way without the need for a specific relativistic program development. However, two-component ECPs which allow to recover the effects of spin-orbit coupling in the heavy atom's valence shell are not in such widespread use.

3
Program Developments for NMR Computations

3.1
The Pre-DFT Era

By the late 1970s a large number of review articles on NMR calculations were already available. They have recently been collected in a meta-review [55]. Most of the theoretical calculations up to that time were concerned with small organic and inorganic main-group systems, while transition-metal systems were generally treated with the help of more or less sophisticated methodologies based on ligand field theory [56]. In 1978 De Brouckere [57] published a 100-page overview with 289 references that was devoted to the linear combination of atomic orbitals SCF MO computation of properties of transition-metal complexes. This included NMR parameters. Both in the case of nuclear shieldings and spin-spin coupling constants De Brouckere found that such calculations had never been carried out for transition-metal systems.

The reader is referred to Ref. [58] for early work on NMR properties based on the Hartree–Fock approach. For instance, in this collection Nakatsuji presented the calculation of shielding constants of transition-metals. It was later concluded that the Hartree–Fock approach is, in general, not adequate for this purpose, though it might be reasonable in the special case of high metal oxidation states or d^{10} systems [1].

3.2
DFT Program Developments for NMR Parameters

During the 1980s, NMR shielding codes were developed based on the Hartree–Fock–Slater (Xα) method, which is now being classified as DFT [59–61]. Computational restrictions did not allow at that time the use of adequate basis sets in order to yield accurate chemical shifts for the small, light atomic, systems that could be studied. For the same reasons, transition-metal complexes were still out of reach. Further, from the results it was concluded that coupled Hartree–Fock is superior to the Xα method [59]. ^{13}C shielding constants were found to have inaccurate magnitudes, though the trends among various small molecules were, according to the authors, reproduced remarkably well [60]. A pioneering finite perturbation theory-Xα study of spin–spin coupling constants in 1976 (only the FC term was implemented) did not yield any agreement with experimental data for H_2 and HF, though reasonable agreement with other theoretical data was obtained.

It is clear that these early DFT attempts suffered from basis set limitations. At the same time, the advantages of having a post-Hartree–Fock method available that is on the one hand potentially accurate enough to treat the NMR properties of transition-metal systems [3,62] and on the other hand computationally feasible (as compared with more accurate but also more expensive ab initio methods), could not be explored. The first modern Kohn–Sham DFT implementation for nuclear shieldings was reported in 1993 [26,63] and employed the IGLO method in order to treat the gauge problem. The LORG approach was implemented in 1994 within the DFT framework [64]. GIAO DFT implementations were soon to follow, see Refs. [25,65–67]. The approach of [67] was also capable of handling hybrid functionals. Results obtained with the CSGT method were also reported. It did not take long before these programs were applied to a vast number of transition-metal systems. The DFT computation of spin-spin coupling constants was initially implemented in 1994 by Malkin et al. [68], followed by Dickson and Ziegler in 1996 [69]. Both used a finite-perturbation approach for the FC term. The SD term was not implemented. The first analytic (nonhybrid) DFT implementation for spin–spin coupling constants (excluding the SD term) appeared only rather recently [45] and was also capable of accurately treating scalar relativistic effects. In fact it had been the huge, variationally unstable, FC matrix elements that did not permit a finite-perturbation treatment for the heavy atomic systems that the program was intended for. An extension of the method in order to include the SD term and spin-orbit coupling was subsequently published [46]. At the same time, two nonrelativistic analytic DFT implementations (including SD) were reported which could additionally treat hybrid functionals as well [70,71]. The B3LYP hybrid functional was shown to yield very promising results for small light atomic molecules, in particular those difficult ones [30] that are not well described by nonhybrid functionals.

3.3 Relativistic DFT NMR Methods

It has been known for a long time [33,72] that relativistic effects can have a pronounced influence on NMR observables. Therefore, relativistic DFT methods were developed soon after the first nonrelativistic codes were available. A significant number of DFT studies, in particular many on transition-metal systems, have focused on relativistic effects on NMR parameters, their origin, and their interpretation. For a detailed account of the methodology development, early and recent applications, and ab initio as well as semiempirical relativistic NMR methods the reader is referred to Refs. [6–8]. The first approach within the DFT framework was to include a scalar relativistic ECP for the metal of a complex when calculating the chemical shift for a ligand nucleus [73,74]. Numerous applications to transition-metal complexes have been carried out using this method (which does not require a relativistic NMR code) [1], some of which will be mentioned later. Schreckenbach and Ziegler have implemented chemical shift computations based on the scalar relativistic Pauli operator and frozen cores. This method could also be applied to metal shifts and was later extended by Wolff and Ziegler [52] to include spin-orbit coupling. The DFT computation of shielding constants based on the two-component ZORA Hamiltonian was implemented by Wolff et al. [43] and Bouten et al. [44]. The author is not aware of DFT computations of NMR observables for transition-metal complexes based on the four-component Dirac or the two-component DKH2 Hamiltonians (Hartree–Fock calculations, mainly on main-group systems but also, with DKH2, on small Hg complexes, have already been reported [47,48,75,76]). Consideration of the Pauli spin-orbit operator in DFT shielding computations as a perturbation, Eq. (11), has been implemented by Malkin and coworkers [77,78] and has been extensively used to analyze spin-orbit effects on nuclear shielding [9]. The incorporation of a spin-orbit ECP into DFT NMR shielding computation was reported by Vaara et al. [79] in 2001. A two-component ZORA DFT approach to compute nuclear spin-spin coupling constants [45,46] was intentionally developed with transition-metal systems in mind because of the known huge influence of relativity in particular on the FC mechanism.

4 Applications to Transition Metal Systems

Before beginning the description of computational results that have been obtained with DFT over the last 10 years, a note on the achievable accuracy is appropriate. Except perhaps for systems that are experimentally very difficult to measure accurately, a quantitative agreement of DFT results with all significant digits that are reported from experiments is virtually never accomplished. The reason for this lies of course in the sensitivity of the NMR parameters not only to experimetal conditions but also to computational approximations. For chemical shifts, the sign is usually determined correctly (except for very small magnitudes). An agreement within 10 – 20% deviation from experiment is usually considered very good, good, satisfactory, or reasonable, depending on the publication and the quality of other available computational data. Similar considerations hold for nuclear spin–spin coupling constants, though the criteria appear to be somewhat less strict because of the difficulties to compute these quantities accurately. Better agreement must in most cases not be expected because the computations are approximate and generally do not account for many influences that affect the experimental value. On the other hand, if there is a very substantial deviation between calculated and experimental results, the present-day DFT methods can usually be trusted to the extent that such a deviation suggests that it is not the functional but the computational model (chemical environment, e.g., solvent, neglect of molecular vibrations) that is deficient. Exceptions are known, of course, for instance the case of ferrocene which will be discussed in Sect. 4.3.

Table 2 Frequently used acronyms. A compilation of references for density functionals can be found, e.g., in the book by Koch and Holthausen [80]

Miscellaneous:	
HF	Hartree-Fock
MP2	Second-order Møller-Plesset treatment for electron correlation
GIAO	Gauge-including (or -independent) atomic orbitals
IGLO	Independent gauge for localized orbitals
AO, MO	Atomic orbital, molecular orbital
HOMO, LUMO	Highest occupied, lowest unoccupied MO
TMS	Tetramethyl-silane
Density Functionals:	
LDA	Local density approximation
GGA	Generalized gradient approximation
XC	Exchange & correlation
VWN	Vosko-Wilk-Nusair LDA
BP or BP86	Becke 1988 Perdew 1986 non-hybrid GGA
LYP	Lee-Yang-Parr non-hybrid GGA
PBE	Perdew-Burke-Ernzerhof non-hybrid GGA
PW91	Perdew-Wang 1991 non-hybrid GGA
B3LYP	3-parameter hybrid GGA
Relativistic approaches:	
ECP	Effective core potential
SO	Spin-orbit, or spin-orbt coupling
pPSC	Perturbational treatment of scalar Pauli operator
vPSC	Variational (frozen core) treatment of scalar Pauli operator
pPSO	Perturbational treatment of Pauli-operator, SO only
pPSCSO	Perturbational treatment of Pauli-operator, scalar + SO
vPSCSO	Dito, but variational (frozen core)
DKH2	Second order Douglas-Kroll-Hess transformation
ZORA	Zeroth-order regular approximation

Some acronyms that are frequently used in the following sections are collected in Table 2. Acronyms referring to individual terms in the nuclear shielding or spin-spin coupling tensor are explained in Table 1.

4.1 Computations of Ligand Nuclear Shieldings in Transition Metal Complexes

In 1995 a large number of ^{13}C chemical shifts in transition-metals were studied with the SOS DFT method (ECPs, PW91 nonhybrid functional), yielding generally good agreement with experimental data. It is also important to note that a hybrid functional (B3LYP) did not perform better in this study than the BP nonhybrid functional. In fact for the MO_4^{n-} systems B3LYP yields results that compare worse with experiment (see later).

Table 3 ^{17}O nuclear shielding constants [$\times(-1)$] in transition metal oxo complexes [a)]

Compound	δ exp. [b)]	BP [c)]	BP [d)]	BP [e)]	B3LYP[d)]	HF [d)]	MP2 [d)]
WO_4^{2-}	129	140	157	138	183	194	21
MoO_4^{2-}	239	216	251	231	289	335	60
CrO_4^{2-}	544	446	508	490	640	1308	−2173
ReO_4^-	278	278	282	277	339	464	−3
TcO_4^-	458	405	421	410	518	819	−184
MnO_4^-	939	778	832	821	1149	7248	−54485
OsO_4	505	521	517	503	657	1295	−1069
RuO_4	815	740	765	733	1037	3330	−8262

a) Experimental data has been converted to absolute ^{17}O shieldings based on a value of 290.9 ppm for $\sigma(^{17}O)$ of liquid water at room temperature [62]. All values in the table have been multiplied by −1

b) As compiled in Ref. 62

c) Scalar relativistic DFT, Pauli operator with frozen core, GIAO. Reference 51

d) Scalar relativistic ECP, GIAO. Reference 62.

e) Scalar relativistic ECP, IGLO, paramagnetic term semi-empirically scaled ("SOS-DFT"). Reference 62.

Experimental and computational results that have been obtained for a number of ^{17}O shifts in tetroxo-metallates MO_4^{n-} are listed in Table 3. It is certainly an intriguing demonstration of the performance of DFT for transition-metal NMR calculations. Compared with the shift range that is covered by these complexes, the errors for all the DFT methods listed are rather small. This is a general finding that has been reported in a large number of studies of ligand shifts in transition-metal complexes [1,3,80]. Relativistic effects for the heavier metals listed in Table 3 have been included either via ECPs or by the vPSC method and, for example reduce the magnitude of the oxygen shielding in WO_4^{2-} by about 50%. It has been argued that spin-orbit effects are of minor importance in this case [1]. The Hartee-Fock method performs badly in comparison. It reproduces the trends, but they are dramatically overestimated. The MP2 results cannot be assigned any physical meaning at all. This is unfortunate since MP2 is at present the perhaps only post-Hartree-Fock ab initio method that would be computationally feasible for NMR calculations for larger transition-metal systems.

The exceptionally strongly shielded protons directly bound to a transition-metal were investigated in Ref. [81]. The calculations were nonrelativistic (except for the Re system, for which the vPSC method was adopted) and employed a frozen core on the metal. Satisfactory agreement with the experimental data was obtained, as shown in Table 4. An analysis of the results in terms of MO contributions revealed that strong, shielding, paramagnetic currents from the adjacent metal fragments cause the large negative chemical shifts. It was found that the magnitude of the proton shielding is determined by the two paramagnetic

Table 4 Experimental and calulated proton chemical shifts (with respect to TMS, all values multiplied by −1) for a number of transition metal hydrides. Data taken from Ruiz-Morales et al., Ref. 81. Nonrelativistic DFT computations (BP functional) except for Re

	$-\delta(^1H)$/ppm	
Complex	calc.	expt.
$[HCr(CO)_5]^-$	6.5	6.9
$[HCr_2(CO)_{10}]^-$	20.3	19.5
$HMn(CO)_5$	6.8	7.5
$HTc(CO)_5$	5.7	
$HRe(CO)_5$ [a)]	5.2	5.7
$H_2Fe(CO)_4$	7.5	11.1
$HCo(CO)_4$	5.3	10.7

[a)] quasi-relativistic calculation

tensor components $\sigma^P_{\|}$ and $\sigma^P_{\perp}$. They are denoted by the orientation of the external magnetic field with respect to the M–H bond, with $\sigma^P_{\perp}$ being the dominant and positive contribution. The induced current strongly reduces the magnitude of the external field component in $\perp$ direction at the position of the hydrogen (i.e., it causes an increased shielding) and is thus responsible for the negative "hydridic" proton shift. The opposite effect is seen for the external field component parallel to the M–H bond. The resulting induced currents reinforce the external field at the proton. It was found that $\sigma^P_{\|}$ is negative, but smaller in magnitude than $\sigma^P_{\perp}$. This effect is sketched in Fig. 1.

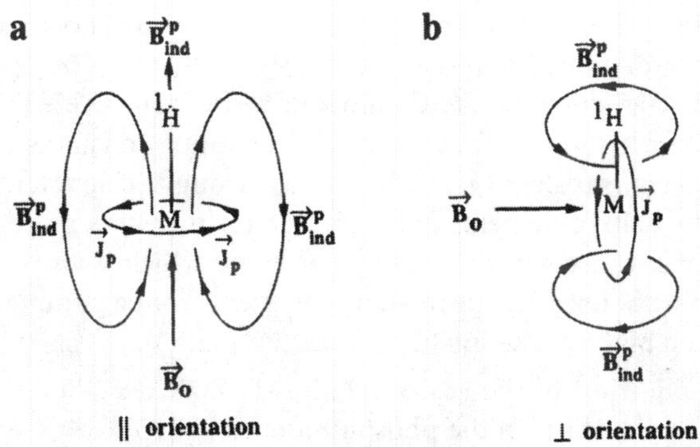

Fig. 1a,b Effect of induced paramagnetic ring currents in low-valent transition-metal hydrides. See also Table 4. (Reprinted with permission from Ruiz-Morales et al. [81], copyright 1996 American Chemical Society)

Table 5 ^{13}C chemical shifts in 5d transition metal carbonyls, from DFT calculations

Compound	δ exp. [a)]	Pauli [b)]	Pauli-SO [c)]	SO-ECP [d)]
$[Hf(CO)_6]^{2-}$	244	228.3	234.5	228.2
$[Ta(CO)_6]^{-}$	211	211.8	214.8	206.0
$W(CO)_6$	192	197.7	196.7	188.6
$[Re(CO)_6]^{+}$	171	183.9	176.2	173.7
$[Os(CO)_6]^{2+}$	147	172.4	149.5	153.9
$[Ir(CO)_6]^{3+}$	121	153.6	125.8	130.9
$[Au(CO)_2]^{+}$	174		165.3	164.0
$[Hg(CO)_2]^{2+}$	169		158.2	154.0

[a)] as compiled in Refs. [52,79], with respect to TMS
[b)] scalar relativistic Pauli operator (variational procedure employing frozen cores), Reference 82
[c)] Pauli operator including spin-orbit coupling (variational procedure employing frozen cores), Reference 52
[d)] Spin-orbit ECPs, Reference 79

Another set of data is compiled in Table 5. Here, the ^{13}C chemical shifts of a series of 5*d* transition–metal carbonyls were calculated with different relativistic DFT approaches. Both scalar relativistic effects as well as spin-orbit coupling contribute to the trend along the series. Again, good agreement with the experimental data is obtained with nonhybrid functionals. It has been shown by Wolff and Ziegler [52] that the observed trend partially results from increasingly large shielding spin–orbit contributions when going from Hf to Ir. At the same time, the HLG gap increases from approximately 2 to 6 eV. According to Eq. (10a) the increasing denominator should lower the magnitude of the spin–orbit contributions. However the binding between the metal and the ligand also changes considerably along the series. While the Hf–C bonds are mostly characterized by $d_\pi - \pi^*_{\mathrm{CO}}$ interactions, for Ir–C it is mostly $d_\sigma - \sigma_{\mathrm{CO}}$. The spin-orbit contributions to nuclear shielding are dominated by the spin–orbit FC mechanism, which is very similar to the one for spin–spin coupling constants [9,33,35], and are more effectively transfered if the bond has strong σ character. In effect, the magnitude of the matrix elements in the numerator increases more strongly than the denominator increases from Hf to Ir. The trend which is already observed at the scalar relativistic level has been shown to stem from a complex interplay of factors, most notably π back–donation from CO [82,83].

A recent application of the scalar relativistic ZORA methodology has been the calculation of ^{31}P shifts in the phosphinidenes [84,85] $\mathrm{Cp}^*(L)M{=}\mathrm{PAr}$, where Cp^* is pentamethylcyclopentadienyl L is a ligand and $M =$ Co, Rh or Ir. Good agreement with experimental data was again obtained with a standard nonhybrid functional (BP). The data are listed in Sect. 4.2, where the calculated ^{31}P–^{31}P coupling constants are discussed. The ZORA approach was also applied by Webster and Hall [86] in order to compute ^{11}B chemical shifts for number of substituted tris(pyrazolyl)borate Rh dicarbonyl complexes. Among other results the study reported a correlation between the computed boron shifts, B–H vibrational stretching frequencies and the denticity of the tris(pyrazolyl)borate ligand.

Complexes with Mo–P and W–P triple bonds were investigated by Wagener and Frenking [87]. ECPs were used for the metals and the BPW91 nonhybrid functional was applied. Both ^{31}P isotropic shifts and anisotropies were calculated and compared with available experimental data. Deviations not exceeding 160 ppm were obtained for $\delta(^{31}\text{P})$, with a studied shift range of about 1000 ppm (largest experimental shift 1364 ppm for $[Mo(P)(N_3N)]$). The signs and the magnitudes of the parallel and perpendicular components of the shielding tensors were found to be in agreement with experimental data for three of the systems, with deviations between theory and experiment being somewhat larger than for the chemical shifts.

A catalytically important system, methylalumoxane, was investigated by Zurek and Ziegler [88]. Computed ^{13}C and proton chemical shifts of its Zr complexes were used to identify the active and dormant species of this "black box" activator of the dimethylcirconocene homogeneous olefin polymerization catalyst by comparison of the NMR parameters of the proposed species with experimental data (BP nonhybrid functional).

DFT computations (BP nonhybrid functional) of ^{13}C chemical shifts were recently used to propose a reaction mechanism that leads to dicubane structures composed of Ti, O, C and alkali metal atoms [89]. ^{13}C shifts of almost 600 ppm were computed and are in excellent agreement with experimental data.

An early application of DFT NMR methods to systems of biological relevance was made by Strohmeier et al. [90]. A combined solid state NMR and computational (BLYP nonhybrid functional) study of several metal 5,10,15,20-tetraphenylporphyrin complexes (Zn, Ni, Mg) achieved good agreement with experimental ^{15}N shielding tensor components. The calculations were also carried out on Zn model systems (phenyl replaced by methyl and H), showing that the ^{15}N NMR parameters were not significantly affected by the substitutions. A stronger effect was found for the shifts of the substituted carbons, thus a comparison of the ^{13}C data for the model systems with experiment was not directly possible. The main influence of the metal on the ^{15}N shielding tensor components was found to be the metal–N distance. Another DFT application to bioinorganic complexes can be found, for example in Ref. [91] (heme proteins and model systems, including ^{57}Fe NMR, B3LYP and BPW91 functionals).

Schreckenbach et al. [92–94] investigated ligand chemical shifts in uranium compounds (ZORA DFT). Nonhybrid ZORA DFT computations were not able to reproduce the trends for ^{19}F shifts of $UF_{6-n}Cl_n$ though the correct magnitudes between 750 and 790 ppm were obtained. Spin–orbit effects were found to be of minor importance for the ^{19}F shifts. ZORA DFT computations were successfully applied to fluorine shifts in the $UF_{5-n}(OCH_3)_n$ series. Similar calculations could also well reproduce the trends for ^{17}O shifts in a number of uranyl complexes. A large-core ECP on uranium in conjunction with the B3LYP hybrid functional did originally reproduce the trends for the ^{19}F shifts in the $UF_{6-n}Cl_n$ series, but the magnitudes were overestimated by 80 - 270 ppm. A recent reinvestigation has shown that small-core ECPs in conjunction with the B3LYP hybrid functional yield results very similar to ZORA PW91 nonhybrid DFT results [95], and the large-core ECP computations have been discredited. It was concluded that the relativistic approximations are not responsible for the failure of the computations to reproduce the experimental trends and it was suggested that the missing treatment of solvent effects might be responsible instead. A chemical shift range of more than 21000 ppm has been predicted for ^{235}U from the computations in Ref. [93]. Unpublished recent calculations by Schreckenbach yielded an even larger chemical shift range of 38000 ppm. However, there are no experimental data available for comparison.

A number of additional DFT applications are reviewed in Refs. [1–3, 96]. In many cases, reasonable results were obtained with nonhybrid density functionals. For 5*d* metal complexes it can be necessary to include spin–orbit coupling, though the trends for ligand shifts are often reproduced already at the scalar relativistic level (similar to the systems in Table 5). Interestingly, it will be demonstrated later that the calculation of metal shielding constants is very sensitive to the choice of the functional and that the hybrid types tend to perform much better, at least for the 3*d* metals. Thus, it is either a very general systematic error compensation, or that the ligand shifts are rather insensitive to the approximations in the most common functionals, which permits us to obtain reasonably accurate ligand shifts for transition-metal complexes.

4.2 Computations of Ligand–Ligand Spin–Spin Coupling Constants

There are at present not very many studies of ligand–ligand coupling constants in transition-metal complexes available. One reason is perhaps that the interest is often in coupling constants between nuclei that are separated by more than one bond, for instance ligands that are both bound to the metal but not to each other. Two-bond coupling constants appear to be more difficult to calculate accurately. Another reason might be that for the DFT calculation of spin–spin coupling constants, generally a coupled perturbed problem has to be solved which results in a higher computational cost than solving an uncoupled problem (as is the case for nuclear shielding with nonhybrid DFT). The number of program codes that have not just the FC term implemented but also the OP, OD, or even the SD term is also smaller compared with the many available codes for nuclear shielding.

Table **6** Experimental and calulated ^{31}P chemical shifts (with respect to PMe_3) and $^2J(^{31}P–^{31}P)$ coupling constants for Cp*(L)M=PAr, with M = Co, Rh, Ir, L = PR_3 or CO, Ar = Mes*. Data taken from Termaten et al., Ref. 85. Scalar ZORA DFT computations (BP functional) on model complexes with L = PH_3 or CO, Ar = H, and Cp instead of Cp* where appropriate

Complex	$\delta(^{31}P)$/ppm		$^2J(^{31}P–^{31}P)$/Hz	
	calc.	expt.	calc.	expt.
(*E*)-Cp(PPh_3)Co=PMes*	903	867	207	76
(*E*)-Cp*(PPh_3)Rh=PMes*	894	868	171	59
(*Z*)-Cp*(PMe_3)Rh=PMes*	942	953	30	19
(*E*)-Cp*(PPh_3)Ir=PMes*	729	687	216	102
(*Z*)-Cp*(PMe_3)Ir=PMes*	768	727	21	18
(*Z*)-Cp(CO)Co=PMes*	1044	1047		
(*Z*)-Cp(CO)Ir=PMes*	844	805		

Among the available examples in the literature, two-bond $^{31}P–^{31}P$ coupling constants ^2J(P–P) have been studied using different methods. Very recently, the scalar relativistic ZORA DFT method has been applied to the calculation of ^2J(P–P) in organometallic systems of the type Cp*(*L*)*M*=PAr, where *M* is Co, Rh or Ir, *L* is PR_3 or CO and Ar is 2,4,6-tri-*tert*-butylphenyl [85]. The computations were based on model systems with *R* = H and Ar = H. The results are listed in Table 6. Though the ^{31}P chemical shifts for the model systems which were already discussed in Sect. 4.1 are in good agreement with the experimental data, the results for the ^2J(P–P) are less favorable. Two trends are reproduced by the computations. First, the ^2J(P–P) for the (E) isomers are consistently larger than for the corresponding (Z) forms. Second, the trend for the magnitude of ^2J(P–P) is Ir > Co > Rh for the (*E*)-Cp*(PR_3)*M*=PAr systems. This is an indication that the influence of the metal is correctly described in the calculations. It is likely that the deviations between the calculations for the model complexes and the experimental data for ^2J(P–P) are caused by electronic effects from the model phosphorus ligands rather than deficiencies in the methodology. Because of the known sensitivity of spin–spin coupling constants to basically all approximations in the computational model it might be necessary to carry out calculations on the same systems that were experimentally studied (and have large bulky ligands on the P's).

$^{31}P–^{31}P$ coupling constants in group VI metal carbonyl phosphines $M(CO)_4(PH_3)_2$, where *M* is Cr, Mo or W, have been investigated by Kaupp et al. [1,97], employing a scalar relativistic ECP on the metal. The results that were obtained with a number of different density functionals are displayed graphically in Fig. 2. It can be seen that there is a significant influence from the specific type of nonhybrid functional that has been applied, testifying to the sensitivity of the properties investigated with respect to the quality of the electronic structure. It can also be seen that the $^2J(^{31}P–^{31}P)$ are noticeably influenced by relativistic effects (about 20 Hz for the trans tungsten complex), with the relativistic computations being closer to experiment than the nonrelativistic ones. The trends for the cis and trans series are well reproduced by this approach and could be related to a few characteristic orbital contributions. The reader is referred to Refs. [35,97] for further details.

Lichtenberger [98] has recently investigated the Si–H coupling constant in the organometallic complex $(\eta^5\text{-}C_5H_5)(CO)_2MnHSiCl_3$. The scalar relativistic ZORA DFT method (BP nonhybrid functional) was employed. A MO-based analysis along with a comparison of calculated and experimental Si–H coupling constants was applied to shed some light on the type of bonding between Si and H and the Mn center and the question whether there is a direct bond between Si and H. For $HSiCl_3$, the experimentally determined magnitude of the Si–H coupling constant

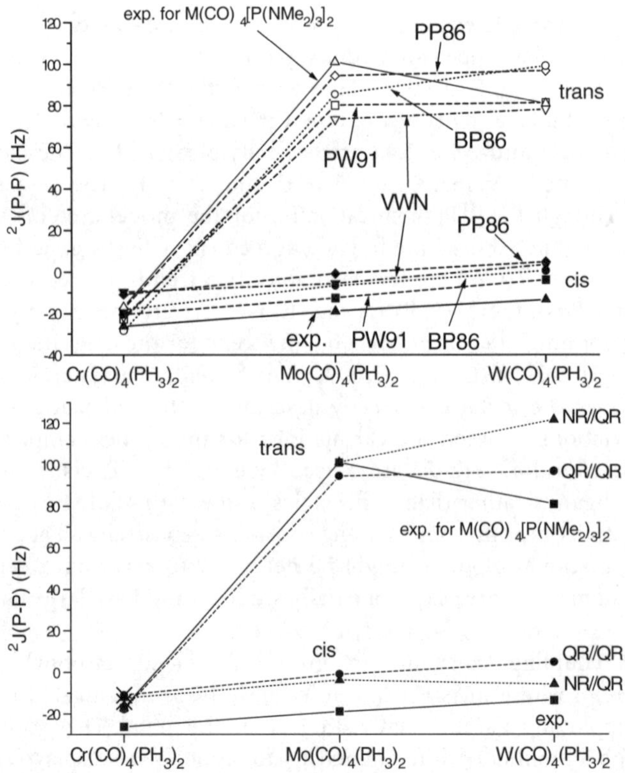

Fig. 2 Influence of the choice of the exchange-correlation potentials and of relativistic effects on the ^{31}P–^{31}P two-bond coupling constants in *cis*- and *trans*-$M(CO)_4(PH_3)_2$, M is Cr, Mo or W. (Graphics courtesy of Kaupp [97])

is 370 Hz (calculated -338 Hz) but is 54.8 Hz in $(\eta^5\text{-}C_5H_5)(CO)_2MnHSiCl_3$ (calculated as -38 Hz for an optimized geometry and -46 Hz when using the experimental Mn–Si distance). The negative sign of the Si–H coupling constant in $HSiCl_3$ is due to the negative magnetogyric ratio of the Si nucleus. In comparison, the Si–H coupling in $(CO)_4Fe(H)(SiCl_3)$ is positive (experiment 21.7, calculated 16 Hz) and is interpreted as a two-bond coupling without a direct Si–H bond contribution. In Ref. [98], the case of $(\eta^5\text{-}C_5H_5)(CO)_2MnHSiCl_3$ was described as an intermediate between these two extremes, in agreement with a 75% reduction of the Si–H bond population in the Mn complex as compared with $HSiCl_3$.

A three-bond $^{13}C\text{–}^{13}C$ coupling constant has been investigated by Autschbach and Ziegler for the complex $[(NC)_5Pt\text{–}Tl(CN)]^-$. This and related complexes will be discussed in more detail in the following sections. On the basis of scalar relativistic ZORA computations (VWN functional), the $C^A\text{–}C^B$ coupling constant was determined as 24 Hz (in the most accurate computational model which included spin-orbit coupling as well as coordination of the Tl center by H_2O). The experimental value [99] is 30 Hz. The three-bond $C^C\text{–}C^B$ coupling, on the other hand, is extremely small. It is noteworthy that the combined contributions due to the SD terms as well as spin-orbit effects on $J(C^A\text{–}C^B)$ in the calculations amounted to as much as 4 Hz (approximately 17% of the total three-bond coupling), resulting from the fact that the coupling is mediated by the strongly delocalized metal-metal and metal-carbon bonds, with a significant influence of relativistic effects in the valence shells of the metals.

A Hartree-Fock and MP2 study of $^1H\text{–}^2H$ (H–D) coupling constants in complexes of the type $[(\eta^2\text{-}H_2)Os^{II}(NH_3)_4L]$, with L being a ligand trans to H_2, by Craw et al. [101] in 1994 did yield magnitudes for the J(H–D) that were similar to the experimental ones, but the trends for different ligands were not reproduced correctly. Large differences were further found between the Hartree-Fock and MP2 values for J(H–D), along with non-systematic sign changes. Only the FC mechanism was considered as a finite perturbation in this work (a coupled Hartree-Fock implementation of the FC term was also reported). A scalar relativistic ECP was used for Os. In 1996, Bacskay et al. [100] reinvestigated the same set of complexes, this time employing DFT (BLYP nonhybrid functional). The agreement between experimental and theoretical data is displayed in Fig. 3 and is much more favorable than what was previously obtained with the Hartree-Fock and MP2 methods. The authors attributed this success to a more accurate description of electron correlation effects at the DFT level as compared with MP2. A strong correlation between the calculated J(H–D) and the strength of the $Os\text{–}H_2$ interaction was found as well as a correlation between J(H–D) and the H–D internuclear distance in the complexes.

4.3
Computations of Metal Nuclear Shieldings

It has turned out that the computation of metal nuclear shieldings and chemical shifts is much more difficult than the calculation of ligand shifts which were discussed previously. It appears that metal shieldings are more sensitive to the quality of the computed electronic structure and consequently larger influences due to the XC potential are observed. Regarding the 3*d* metals, it has turned out that hybrid functionals appear to be particularly well suited for NMR computations of the metal shielding constants. This cannot be easily extrapolated to all of the transition–metals, though, since counter examples are known for which nonhybrid functionals perform better. On the other hand, some 4*d* metals have been treated most successfully with hybrid functionals.

Calculated versus experimentally observed ^{103}Rh chemical shifts for 13 complexes are displayed in Fig. 4 [3]. The nonrelativistic DFT computations (B3LYP hybrid functional) were obviously able to achieve excellent agreement with experiment. In comparison, a nonhybrid functional also achieved a linear relation

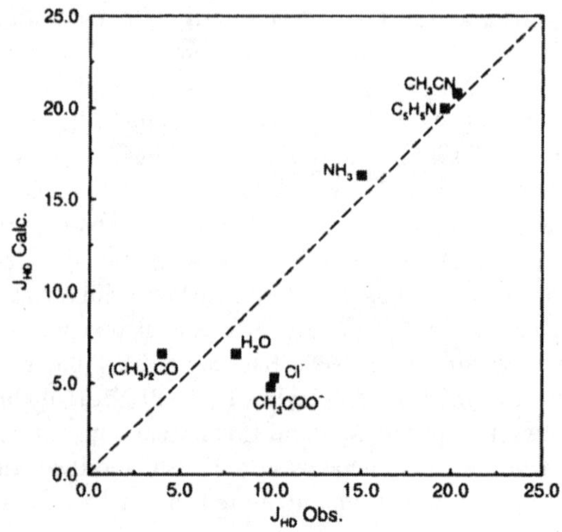

Fig. 3 Calculated verrsus experimental ^{1}H–^{2}H (H–D) spin–spin coupling constants in the complexes $[(\eta^2\text{-H}_2)\text{Os}^{\text{II}}(\text{NH}_3)_4L]$, from density functional theory (DFT) calculations (BLYP). The ligand L is indicated in the graphic. The *dashed line* does not represent a fit but indicates where J_{HD}Calc. = J_{HD}Obs. (Reprinted with permission from Ref. [100], copyright 1994 American Chemical Society)

between the calculated and the experimental values, but the slope of the linear fit was 0.9 instead [1, 102]. ^{99}Ru chemical shifts have also been calculated more accurately with the B3LYP functional [103]. It has further been concluded from the success of these and other nonrelativistic computations on 4*d* metal shifts [1, 3] that relativistic effects on the shielding tensor, though known to be quite substantial, almost completely cancel when the chemical shifts are evaluated. These findings indicate that (at least the scalar) relativistic effects on shielding constants for 4*d* (and lighter) metals almost exclusively originate from contributions from the core orbitals, which are not substantially altered by the chemical environment of the metal. It should be noted that these considerations apply if the metal is the only heavy atom in the complex. If there are heavy ligand atoms present, in particular the spin–orbit effects on the metal shift originating from a heavy neighboring atom should be considered. In a Hartree–Fock study of neighboring-atom-induced-spin–orbit contributions to Nb and Ti chemical shifts it was argued, on the other hand, that this spin–orbit effect is canceled by competing effects in the metal's valence shell [104], which appears to be rather typical for early transition-metals. For 5*d* metals the situation is, in general, different and relativistic effects on the metal shielding can be induced by the metal itself. This point will be discussed further later.

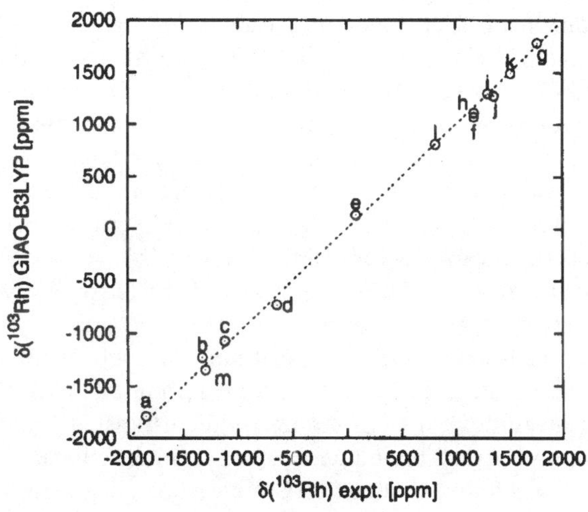

Fig. 4 Calculated versus experimental ^{103}Rh shifts for a number of Rh complexes. (Reprinted with permission from Bühl et al. Ref. [3], copyright 1999 John Wiley & Sons)

It is a typical observation that for many 4*d* and 5*d* metals nonhybrid functionals perform as well as, and sometimes better than, hybrid functionals. In the case of ^{95}Mo chemical shifts, the nonhybrid BPW91 functional yields superior results to B3LYP [3]. Bryce and Wasylishen have recently found excellent agreement for the ^{95}Mo (and ^{13}C) chemical shift with solid-state NMR data for the mesitylenetricarbonylmolybdenum(0) piano-stool complex. The ZORA DFT calculations (BP nonhybrid) functional yielded -1873 ppm, as compared with the experimental value of -1885 ± 2 ppm. A nonrelativistic DFT calculation employing the B3LYP hybrid functional yielded -460 ppm instead. It is thus of importance to investigate carefully the performance of various functionals before drawing general conclusions from DFT computations on 4*d* metal chemical shifts.

For the 3*d* metals there is now much evidence showing that nonhybrid functionals are not very reliable for metal chemical shift computations, but that B3LYP and other hybrid functionals perform very well. In 1997 Bühl [102] investigated (along with Rh) the illustrative case of ^{57}Fe chemical shifts. The slopes of linear fits of the calculated to the experimental data for nine iron complexes were 0.65 for the nonhybrid BPW91 functional, but an excellent 0.97 was obtained with the B3LYP hybrid functional. The BPW91 results for ferrocene were particularly poor (657 ppm versus an experimental value of 1532 ppm), whereas B3LYP achieved almost perfect agreement with experiment (1485 ppm or 1525 ppm, depending on the basis set). Similar results were recently reported for eleven ^{55}Mn chemical shifts [105] (slopes of 0.72 and 0.99, respectively). Pioneering nonrelativistic DFT studies of ^{59}Co chemical shifts have also concluded that the B3LYP and other hybrid functionals are able to achieve good agreement with experiment [106, 107], but that the nonhybrid BLYP functional leads to less satisfactory results [106]. On the other hand, the first systematic DFT study of transition-metal shifts concluded that $\delta(^{51}\mathrm{V})$ is equally well described by hybrid and nonhybrid functionals [108], with mean errors of about 3% of the investigated chemical shift range of 3500 ppm. The performance of the B3LYP functional for a large number of calculated 3*d* and 4*d* metal shifts, including some of the previously discussed cases is displayed in Fig. 5. Except for Mo, the overall performance of the hybrid DFT calculations is very satisfactory.

The infamous case of the ^{57}Fe chemical shift in ferrocene was later studied in greater detail by Schreckenbach [51]. Compared with the reference (iron pentacarbonyl) ^{57}Fe in ferrocene is shielded more strongly owing to the smaller magnitude of the negative paramagnetic contribution σ^{P}. With the nonhybrid functionals the magnitude of σ^{P} is strongly underestimated, which results in a too small a chemical shift. The influence of a hybrid functional is threefold. On the one hand, it was found that the HLG approximately doubled owing to the admixture of Hartree–Fock exchange in the XC potential. However, according to Eq. (9a) one would naively expect that this causes a reduction of the magnitude of σ^{P} and thus an even smaller chemical shift. At the same time, the hybrid functional causes the orbital shapes to alter significantly, which was found to result in much larger matrix elements in the numerator of the paramagnetic term in Eq. (9a). A third factor is the coupled contribution from Hartree-Fock exchange, which was also found to increase the magnitude of σ^{P}. The previously mentioned SOS DFT approach only affects the energy denominators in Eq. (9a) and was consequently found to yield an even smaller chemical shift for ferrocene than a standard uncoupled method [102] (133 ppm, experiment 1532 ppm). The findings by Schreckenbach are probably transferable to other 3*d* transition-metal systems.

^{113}Cd chemical shifts have recently been calculated for a number of bioinorganic complexes [109, 110]. The nonrelativistic DFT method employed the B3PW91 hybrid functional. The agreement with experimental data from solid state NMR was satisfactory (deviations of less than 30 ppm on a shift range of about 400 ppm) for most of the complexes. For Cd alaninate and Cd glycylglycinate, the differences were larger (74 and 48 ppm, respectively). Individual tensor components were also studied and they compared well with experiment. An earlier computational investigation of Cd chemical shifts was carried out on

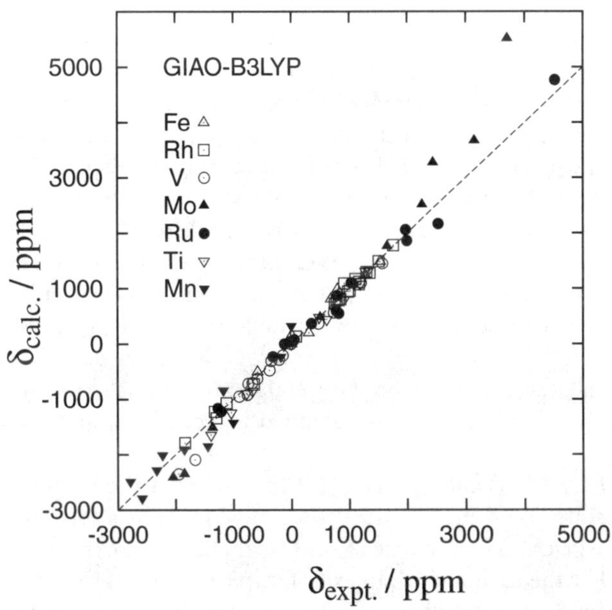

Fig. 5 DFT computed metal NMR chemical shifts (B3LYP hybrid functional) for a large number of inorganic and organometallic transition-metal compounds. The *dashed line* is not a fit but indicates where $\delta_{\text{calc}} = \delta_{\text{expt.}}$. (Graphics courtesy of Bühl [5])

model systems for the active site of the Cd(II)-substituted enzyme phosphotriesterase [111] (B3LYP hybrid functional). The computed pronounced dependence of the Cd shielding constant on the metal's coordination number was used in conjunction with information from X-ray diffraction data to determine which of a number of computed equilibrium model structures of similar energy is likely to correspond to the enzyme's structure in solution.

For further data on DFT computations of 3*d* and 4*d* metal shieldings the reader is referred to Refs. [1,3,5].

When 5*d* metals are to be included in a study of metal shifts a relativistic NMR approach should be used. There are cases in which the heavy atom's (HA) effects on its own shift (this has been coined a heavy atom–heavy atom effect [112]) approximately cancel, but this cannot always be expected. The reason for this lies in the fact that for the heavy 5*d* metals relativistic effects have a very substantial influence on the way they bind to their ligands [37] (along with bond-length contractions of up to several 10 pm). This alters qualitative features of the bond, for instance, the metal 6*s*/5*d* contributions in metal–ligand bonds. Since the chemical shift with respect to a reference compound results precisely from the different bonding environment, relativity will have an effect on the chemical shift as soon as its acts strongly — and noticeably differently — in the metal-ligand bonds of the reference and the probe. This is not so much the case for the 4*d* metals for which relativistic effects on the chemical bond are much smaller and thus the most prevalent relativistic contributions are from the core shells that cancel between probe and reference. As an example, ^{199}Hg shifts of Me–Hg–X, where X is Cl, Br or I, with respect to the $HgMe_2$ reference, excluding or including spin–orbit contributions in the shielding constants, are displayed in Fig. 6. The DFT results are in reasonable agreement with experiment, though the trend is not well reproduced. In comparison, the Hartree–Fock results, though close to the DFT values when spin–orbit coupling is excluded, yield a much too large chemical shift range for the series. It is obvious, though, from both DFT and Hartree–Fock computations that there are very considerable spin–orbit relativistic effects on the ^{199}Hg chemical shift that need to be properly described in a computation. The scalar relativistic contributions to δ(Hg) are also substantial. A larger set of computed Hg chemical shifts is compiled in Refs. [6,7] and allows the same conclusion.

Scalar relativistic (Pauli operator, BP functional) frozen-core computations of the metal shifts in group VI oxides MO_4^{2-} and carbonyls $M(CO)_6$ showed that scalar relativistic effects on the metal shift strongly increase from the 3*d* to the 5*d* metals [51]. For the carbonyl complexes, for instance, $\delta(^{95}\text{Mo})$ was only reduced by 10 ppm owing to relativistic effects, whereas the corresponding change in $\delta(^{183}\text{W})$ was found to be -435 ppm, leading to a much better agreement with experiment. These and further computational data on ^{183}W chemical shifts are collected in Table 7. Some Hartree–Fock data are also listed for comparison. The

Table 7 ^{183}W chemical shifts, from DFT and Hartree–Fock computations

Compound	δ exp. [a]	DFT vPSC [51]		DFT vPSCSO [113]	DFT ZORA [113]	HF DKH2 + vPSO [114] [b]	
		nrel	rel	rel	rel	nrel	rel
$W(CO)_6$	-3446	-4075	-3703	-3306	-3876		
WF_6	-1121			-107	-630	-1795	-1135
WCl_6	2181			1773	1932	2266	2686

[a] with respect to WO_4^{2-}, as compiled in Ref. [113]

[b] Scalar relativistic DKH2 approach with Pauli spin–orbit operator, reported in [114] with respect to WF_6

metal chemical shifts of the tetroxo-metallates listed in Table 3 have been studied mainly with relativistic effects in mind. The DFT ZORA and vPSCSO methods were compared and ZORA has emerged as the more accurate one. Unfortunately, the oxygen shieldings were not reported for comparison.

A larger number of tungsten chemical shifts were computed with the relativistic ZORA and the vPSCSO approaches in Ref. [113]. Pb shifts were also studied in this work. A rather small mean error of 3% of the investigated ^{183}W

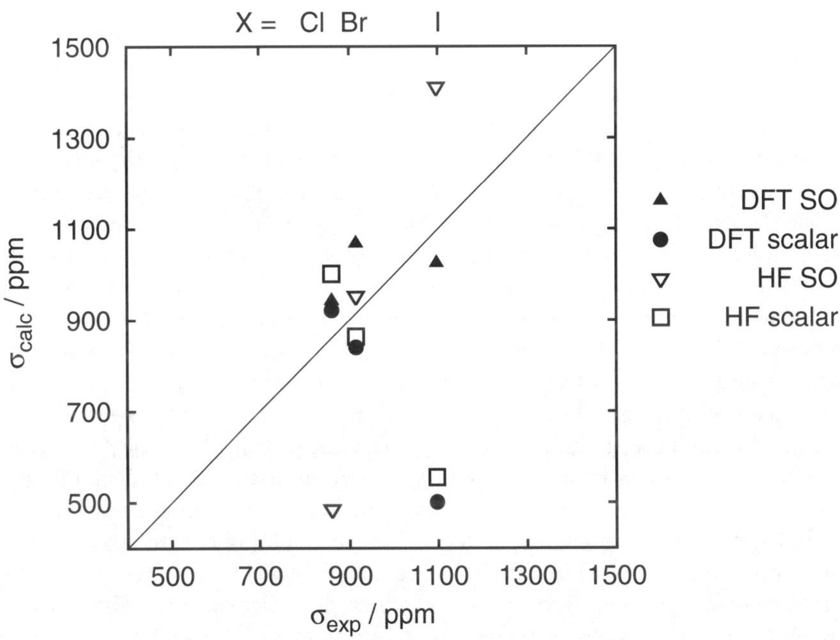

Fig. 6 ^{199}Hg chemical shifts (with respect to $HgMe_2$) for MeHg*X*, *X* is Cl, Br or I, from zeroth-order regular approximation *(ZORA)* DFT (Perdew–Wang nonhybrid generalized gradient approximation functional) and second-order Doudlas–Kroll–Hess Hartree–Fock (HF) computations. Data taken from Refs. [43,48]

chemical shift range of 7000 ppm was obtained from the ZORA computations (the Pauli approach yielded 6% error). Using the same methodology, the ^{183}W nucleus in the exotic species WAu_{12} has recently been predicted to be exceptionally strongly shielded, resulting in a chemical shift of about -13000 ppm [115]. No experimental data are yet available in order to confirm this finding.

The ZORA DFT approach was further applied by Gilbert and Ziegler [116] to the computation of ^{195} Pt chemical shifts. The accuracy of the result was less favorable, with a mean error of about 10% of the investigated chemical shift range. A detailed analysis of the result by means of the HLG and the magnitude of the contributing matrix elements was carried out. As in previously mentioned cases, the HLG alone would cause opposite trends for δ(Pt) along a series of complexes (e.g., for a series of Cl, Br and I ligands), but is overcompensated by changes in the matrix elements (the numerator of Eq. 9a).

The reader is referred to Refs. [6,7] for further examples of relativistic NMR metal shieldings computations.

Now that computationally efficient DFT methods for metal chemical shifts are available and established, applications have started to appear in which comparisons of computational results for isolated metal complexes with experimental (usually solution) data are questioned and more elaborate approaches are taken. With the known sensitivity of the shielding tensor in mind, it is natural to include effects from solvation into the computations, and to consider vibrational effects and temperature. Applications to transition-metal chemical shifts have been carried out by Bühl and coworkers [12–14,96,117]. The protocol consisted of performing ab initio molecular dynamics [118] simulations (Car–Parrinello or Born–Oppenheimer type) of the complexes at a given temperature, including solvent molecules where necessary, and averaging NMR computational results (DFT, B3LYP hybrid functional) for a number of snapshots along the trajectory. For vanadate complexes in aqueous solution [96,117] and the permanganate ion [12] differences between the averaged chemical shift δ_{av} and the one obtained from a single computation at the equilibrium structure δ_e did not exceed 50 ppm, which is rather insignificant when compared with the chemical shift ranges for these metals. On the other hand, for the ^{57}Fe chemical shift in $[Fe(CN)_6]^{4-}$, δ_{av} with respect to iron pentacarbonyl was found to be more than 1500 ppm smaller than $\delta_e = 4120$ ppm, on the basis of simulations including water molecules [14]. This included a -246-ppm change of the reference shielding constant due to thermal averaging. The experimental chemical shift is 2455 ppm. Without adding the solvent, δ_{av} was even larger than δ_e, which thus increased the difference between the computed and the experimental value by another 455 ppm. The final result of $\delta_{av} = 2593$ ppm based on the simulations with solvent is in excellent agreement with the experimental value of 2455 ppm when taking the enormous thermal and solvent effects into consideration. These effects were rationalized by the exceptional sensitivity of $\delta(^{57}Fe)$ to the metal–ligand bond distances and the high charge of the complex. The study was later extended to investigate the strong shielding of ^{57}Fe ($\delta = 2004$ ppm) in $[Fe(CN)_5NO]^{2-}$ compared with $[Fe(CN)_6]^{4-}$ ($\delta = 2455$ ppm) [13]. For the former, the calculated δ_{av} was found to be in good agreement with experiment (about 1997 and 2076 ppm upon inclusion of solvent, with the Car–Parrinello and Born–Oppenheimer simulations, respectively) and differed from δ_e by about 250 ppm.

Solvent effects on ^{199}Hg chemical shifts were investigated computationally by Wolff et al. [43] as early as 1999. Here, explicit solvent molecules were included in relativistic ZORA DFT NMR computations (PW91 nonhybrid functional) in order to account for the direct metal–solvent interaction occurring in solution. The computations could correctly reproduce the sign and the magnitude of solvent changes of approximately 10^3 ppm which had already been known from NMR measurements for a long time. The geometry change (bending) of linear HgX_2 due to solvent complexation was found to be of similar importance as the direct electronic effect of the solvent.

4.4 Computations of Metal–Ligand and Metal–Metal Spin–Spin Coupling Constants

When discussing computational results for spin–spin coupling constants it should be kept in mind that this molecular property is at least as, and often even more, sensitive to the approximations in the computational model as nuclear shieldings (geometries, basis sets, XC potentials, environment, etc.). Thus, an agreement with experiment within some 10% margin must be classified as excellent for a molecule the size of a typical (even small) transition-metal complex. Errors of 20% are often still considered acceptable since for more difficult cases it is sometimes even a challenge to calculate the correct sign of the coupling constant. In addition, relativistic effects can be enormous for 5*d* metals and significant even for the 3*d* metals. This is among the reasons why many DFT studies of metal–ligand and metal–metal coupling constants have focussed on relativistic effects.

First-principles theoretical metal–ligand one-bond coupling constants for 3*d* metal complexes were first reported as part of the validation study of Dickson and Ziegler's nonrelativistic DFT implementation [69] (VWN nonhybrid local functional). The reduced *M*–C coupling constants for the carbonyls $[V(CO)_6]^-$, $Fe(CO)_5$ and $[Co(CO)_4]^-$ were calculated as 127, 213 and 353 $\times 10^{19}$ T^2J^{-1}, as compared with experimental values of 146, 239 and 400$\pm$20 T^2J^{-1}, respectively (averaged value for equatorial and axial carbons for the iron carbonyl). The approximately 15% underestimation of the coupling constants was attributed to approximations in the density functionals, the missing SD mechanism, basis set limitations, geometries, neglected vibrational effects, or a combination of all these. It is also likely that missing relativistic corrections on the *M*–*L* coupling are of the order of 10% and thus might account for most of the deviations between theory and experiment.

A larger set of samples was later studied by Khandogin and Ziegler [119], employing the same program code (BP nonhybrid functional). The results are collected in Table 8. It can be seen that the coupling constants for the complexes $[V(CO)_6]^-$, $Fe(CO)_5$ and $[Co(CO)_4]^-$, are only slightly affected by the different choice of functional, since the 1996 results of Dickson and Ziegler are almost exactly reproduced.[2] It should be noted that differences due to the choice of the XC potential are usually somewhat larger because of the sensitivity of nuclear spin–spin coupling constants.

Owing to the fact that the FC mechanism was implemented in form of a finite perturbation, the ligand nucleus had to be chosen as the "perturber"; otherwise the procedure would not yield reliable results. (In a properly implemented analytical scheme the result does not / must not depend on which nucleus is chosen as the perturbing one.) The OP term was treated in an uncoupled manner, similar to the paramagnetic portion of σ_A in nonhybrid DFT schemes. It was found that this approach resulted in reasonable agreement with experiment in almost all cases, with permanganate being the notable exception. This complex was found to be difficult because the small magnitude of the coupling constant is caused by a large, positive FC term which is almost completely canceled by a negative OP contribution. Accordingly, the relative error in the calculation is large. Substantial OP contributions were also seen for the M–F coupling constants. Generally, when one (or both) of the coupled atoms has many lone pairs, the OP term in K_{AB} can be expected to be large. For instance, it dominates the coupling constants for diatomics of the p-block of the periodic table [46, 120].

Khandogin and Ziegler performed an extensive analysis of the computational results for the FC mechanism. It was found that only MOs with s-type contributions from both the metal and the ligand contribute to the FC portion of $K(M\text{–}L)$. This can in turn be interpreted as the spin–spin coupling constant being a valence property, since core orbitals do not have large values at both nuclei. Lone-pair orbitals hardly contribute directly for the same reason, but they can have important indirect effects, as noted earlier. Some emphasis was placed on relativistic effects, since the nonrelativistic results for some of the 5d metal complexes drastically deviated form the experimental data. See, for instance, the entries for WF_6, $W(CO)_6$ and $Pt(PF_3)_4$ in Table 8. The scalar relativistic frozen-core computations (along with nonrelativistic operators implemented in the spin–spin coupling code) did not yield a sufficient improvement over the nonrelativistic results (except for WF_6) and it was concluded that a more so-

[2] It is not explicitly stated in Ref. [69] which functional was used; however, it is noted that for the set of small molecules the "LDA" results compare well with theoretical data in the literature. It is thus reasonable to assume that the VWN functional was employed throughout

Table 8 Reduced one-bond metal-ligand spin-spin coupling constants for some transition metal complexes, in $10^{19}T^2J^{-1}$. DFT computations (BP non-hybrid functional) employing all-electron Slater-type basis sets for the 3d metal M-C and M-O couplings, Slater-type frozen-core basis sets for the metals in all other cases. Data taken from Khandogin & Ziegler, Ref. 119

Complex	Coupling K	calc.[a]	expt.
$V(CO)_6^-$	V–C	129	146
$Fe(CO)_5$	Fe–C [b]	216	239
$Co(CO)_4^-$	Co–C	348	400±20
$Nb(CO)_6^-$	Nb–C	226/262	319
$Mo(CO)_6$	Mo–C	251/293	344
$Rh(CO)_4^-$	Rh–C	604/713	778
$W(CO)_6$	W–C	547/816	997
VO_4^{3-}	V–O	150	144
CrO_4^{2-}	Cr–O	104	108
MnO_4^-	Mn–O	29	75
MoO_4^{2-}	Mo–O	355/346	380
TcO_4^-	Tc–O	408/403	350
ScF_6^{3-}	Sc–F	49	59.6
TiF_6^{2-}	Ti–F	51	53.2
VF_6^-	V–F	−27	29.7
NbF_6^-	Nb–F	96/74	125
WF_6	W–F	260/84	86.6
$V(PF_3)_6^-$	V–P	466	399
$Cr(PF_3)_6$	Cr–P	460	386
$Co(PF_3)_4^-$	Co–P	861	1063
$Ni(PF_3)_4$	Ni–P	903	1080
$Nb(PF_3)_6^-$	Nb–P	798/978	883
$Mo(PF_3)_6$	Mo–P	780/948	879
$Pt(PF_3)_4$	Pt–P	2981/3770	6215

[a] Nonrelativistic/scalar-relativistic (vPSC) DFT calculations for the 4d and 5d metals (nonrelativistic spin-spin coupling perturbation operators). Nonrelativistic results for 3d metals

[b] Average calculated value for equatorial and axial carbons ((3 eq. + 2 ax.)/5)

phisticated treatment is necessary. It was also shown that, though the AO matrix elements in particular of the FC operator always increased relativistically, the relativistic effect on the coupling constant might be positive or negative because of different signs of individual contributions. The same authors subsequently tried to extend the frozen-core scalar relativistic Pauli approach in order to yield improved relativistic values for the FC matrix elements [53]. This was tested on a number of 5*d* metal complexes, including WF_6, $W(CO)_6$ and $Pt(PF_3)_4$, but with limited success regarding the agreement with experiment. This work also contains a study of the trans effect in square-planar Pt complexes and its relation to the one-bond Pt–P coupling constants.

Significant progress in the field of relativistic DFT spin–spin coupling computations involving one heavy metal nucleus, or even two metal nuclei, has been made recently due to the availability of the ZORA DFT approach [45,46]. For the late 3*d* metals, relativistic effects can roughly be estimated to be of the order of 10% for a metal–ligand coupling [121]. For the 5*d* metals they might amount to 100% or more of a respective nonrelativistic result. Up to an order of magnitude increase might be expected for a coupling constant between two 5*d* metals. These estimates are based on relativistic scaling factors for the FC term [121, 122] and serve only as an order-of-magnitude guideline, and only if the coupling constants are not dominated by one of the other mechanisms (most notably OP). A number of computations have so far demonstrated that relativistic correction factors for the FC mechanism are of limited use. In particular for 5*d* metals a scaling of nonrelativistic AO contributions to the FC mechanism does often not yield correct results as the relativistic effects on the FC matrix elements might be quenched or amplified by relativistic effects on the chemical bonds.

A comparison of scalar ZORA DFT results (VWN nonhybrid local functional) with experimental data for the one-bond metal–ligand coupling constants of a number of complexes containing W, Pt and Hg (a few data points for Pb are also included) is displayed in Fig. 7.

The result for $Pt(PF_3)_4$ is indicated and should be compared with the nonrelativistic value in Table 8. An increase of more than 100% of the nonrelativistic value brings the computed result in excellent agreement with experiment [45]. The case of $Hg(CN)_2$ is similar, with the nonrelativistic Hg–C coupling being about 40% of the experimental value. However, the scalar ZORA DFT computation on $Hg(CN)_2$ yields "only" 443 $\times 10^{20}$ T^2J^{-1}, but the experimental value in methanol is 577.8 T^2J^{-1}. Spin–orbit corrections are not responsible for the deviation (they amount to about -18×10^{20} T^2J^{-1}). It was found that up to four methanol molecules had to be included in the computations [124]. Their coordination to the Hg center increased the coupling constant by another 135 $\times 10^{20}$ T^2J^{-1}, yielding almost exactly the experimental value. It was found that such direct solvent coordination resolved most of the deviations between calculated and experimental values that had remained in the original study, Ref. [45]. Thereby, only complexes with open coordination sites have to be considered, such as the linear Hg and the square-planar Pt complexes. An analysis of the effect [124] indicated that it might be caused by a small amount of charge donation from the solvent into the metal–ligand bonds. The effect is stronger the more nucleophilic the solvent is and the more solvent molecules are coordinated to the heavy metal. Generally, it was found that upon completion of the first coordination sphere of the metal by solvent molecules, agreement with experiment becomes at least satisfactory, often excellent. Even "inert" solvents such as chloroform can cause a substantial solvent effect.

An even more pronounced example of solvent effects on spin–spin coupling constants is seen to govern the NMR spectrum of the complexes $[(NC)_5Pt\text{–}Tl(CN)_n]^{n-}$, $n = 0, 1, 2, 3$ (I–IV) and the related system $[(NC)_5Pt\text{–}Tl\text{–}Pt(CN)_5]^{3-}$ (Fig. 8).

A preliminary computational ZORA DFT investigation indicated that at least 50% of the huge magnitude of the Pt–Tl coupling constant (experiment 57 kHz) of the $n = 1$ system II is due to coordination of Tl by solvent molecules (water) [125]. This was also the first DFT study of a coupling constant between two heavy metals. A recent study of the whole series has confirmed the large solvent effects and shown that their influence on the $n = 0$ system I is even larger [126]. Thereby, it was also found that bulk solvent effects need to be included in the computations as well, and that the simultaneous application of both a continuum solvent model (conductor-like scrrening model [127, 128]) and an

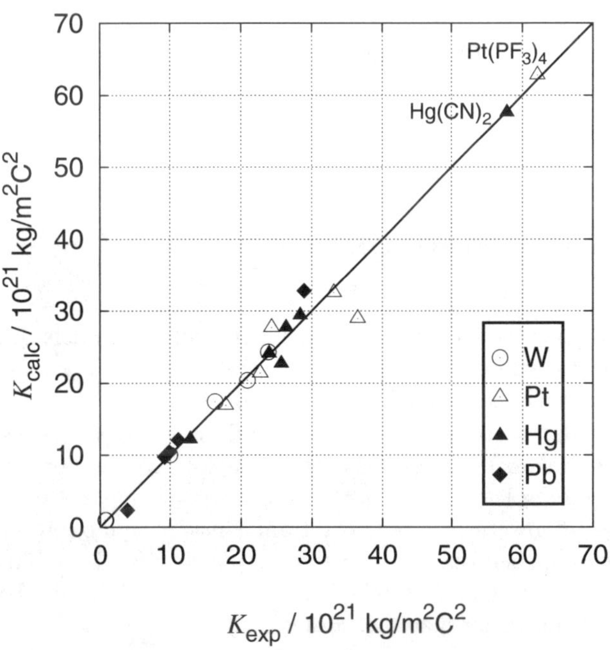

Fig. 7 One-bond metal–ligand reduced nuclear spin–spin coupling constants (absolute values) for 5*d* metal complexes. Scalar relativistic ZORA DFT results (*VWN* functional) versus experimental values. A few data points for Pb are also included. Data taken from Refs. [45,123]. The *line* is not a fit but indicates where $K_{calc} = K_{exp}$. The different *markers* indicate the different metals. For not octahedrally or not tetrahedrally coordinated metal centers, the first coordination shell was completed with solvent molecules, as described in Ref. [124]

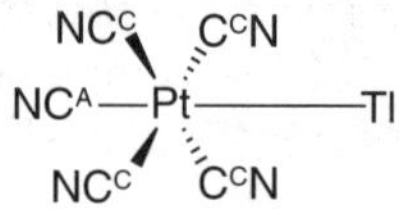

I

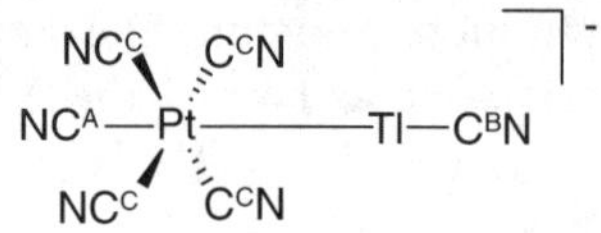

II

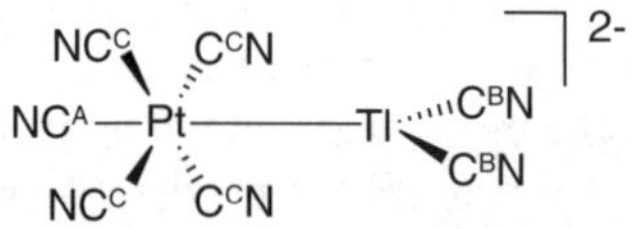

III

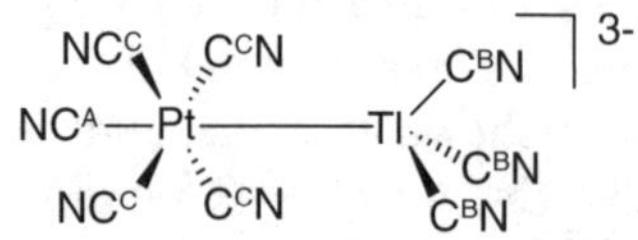

IV

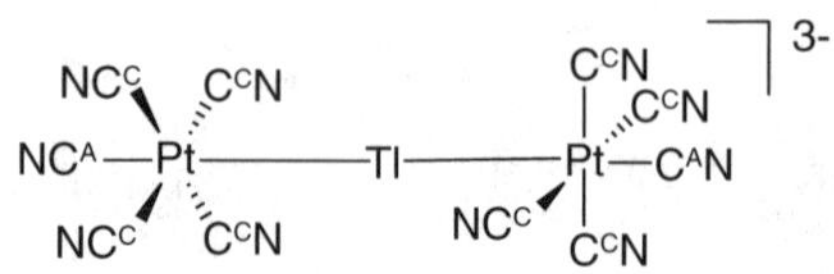

V

Fig. 8 Structures of complexes I–V. See text for details

improved XC potential (statistical averaging of orbital potentials) [32,129] yields good agreement with the experimental values for the Pt–Tl coupling constant. The computational results are displayed in Fig. 9.

The metal–ligand coupling constants for these systems were also studied in Ref. [126], with similar success. The Tl–C coupling constants are very strongly influenced by the solvent, whereas the Pt–C coupling constants are less affected though some improvement is obtained from considering the solvent. Solvent coordination was further found to be mainly responsible for the unintuitive relative magnitudes of ${}^{1}J$(Tl–C) and ${}^{2}J$(Tl–C) for complex II [125], with the latter being more than 3 times larger than the former.

The magnitudes of ^{195}Pt–^{195}Pt coupling constants were the subject of a theoretical investigation in Ref. [131]. It has long been known experimentally that Pt–Pt coupling constants in chemically closely related dinuclear complexes can differ by an order of magnitude — with no obvious correlation to Pt–Pt internuclear distances [132, 133]. Representative examples are $[\{Pt(CO)_3\}_2]^{2+}$, $[\{PtCl(CO)(PPh_3)\}_2]$ and $[\{Pt(CNCH_3)_3\}_2]^{2+}$, with small Pt-Pt coupling constants of 551, 760 and 507 Hz, respectively, and on the other hand

$[\{PtCl_2(CO)\}_2]^{2-}$ and $[\{PtBr_2(CO)\}_2]^{2-}$, with large Pt-Pt coupling constants of 5250 and 4770 Hz, respectively. ZORA DFT (VWN local nonhybrid functional) computations could reproduce this order-of-magnitude difference for the systems $[Pt_2(CO)_6]^{2+}$ and $[Pt_2(CO)_2Cl_4]^{2-}$. An analysis revealed that solvent effects are not responsible for these trends, but they are caused by the different strength of the σ interaction of the ligands CO and Cl^- with the metal–metal fragment. A simple argument based on Hückel theory shows why a stronger interaction, in particular in an axial position, causes a reduction of the magnitude of J(Pt–Pt). The trend is greatly amplified by relativistic effects, resulting in the mentioned difference of an order of magnitude.

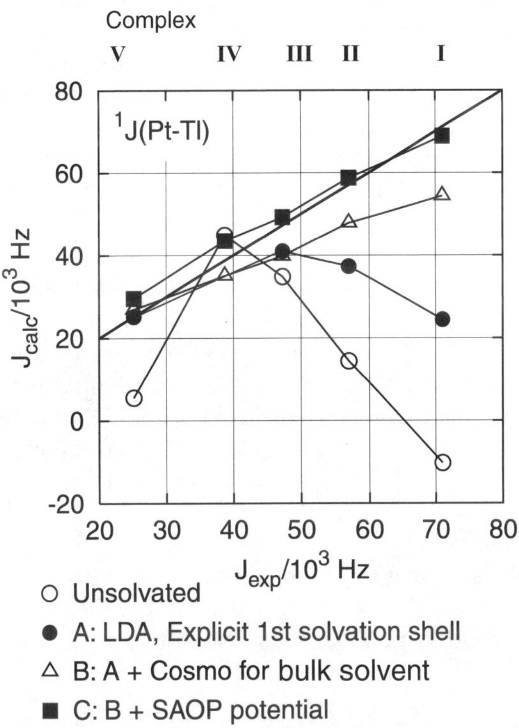

Fig. 9 Nuclear spin-spin coupling constants $J(^{195}\text{Pt}-^{205}\text{Tl})$ for complexes I–V (see Fig. 8), from ZORA DFT computations. Data taken from Autschbach and Le Guennic [126]. Different computational models were applied: Model *A* includes explicit water molecules. In Model *B*, a continuum model (conductor-like screening model, *COSMO*) is applied in addition to the explicit solvent molecules of model A. Model *C* differs from model B in that instead of the VWN functional the statistical averaging of orbital potentials (SAOP) XC potential was used, which allows more accurate computations of NMR parameters [32]. The NMR measurements were carried out in aqueous solution [99, 130]

In Ref. [134] the extremely large coupling constants between ^{199}Hg centers in complexes containing Hg_2^{2+} and Hg_3^{2+} were investigated. It was found that consideration of coordination and polarization of a metal–metal fragment can rationalize the magnitude of the metal–metal coupling constants. The coordination can thereby be due to ligands or solvent molecules. Unlike some previously studied systems in which only one metal center could be coordinated by the solvent [124,125], solvent (or ligand) coordination of both metals of a metal–metal fragment reduces its coupling constant. A strongly polarized metal–metal bond also exhibits smaller couplings. Both effects could be rationalized by intuitive arguments based on Hückel theory [134]. For the complex of Hg_2^{2+} with the crown ethers 18-crown-6 and 15-crown-5, very good agreement with the experimental ^1J(Hg-Hg) of 284 kHz (^{199}Hg) [135] could be achieved. This is the largest currently experimentally determined coupling constant. The calculated scalar ZORA DFT result is 278 kHz (VWN local nonhybrid functional), missing spin-orbit corrections were estimated to be of the order of -10% or smaller. At the same time, the coupling constant of the (hypothetical) free Hg_2^{2+} is likely to be around 0.9 MHz, indicating that an upper limit for Hg–Hg coupling constants has not yet been reached. However, a system such as the Hg_2^{2+}(18-crown-6)$_2$ complex for which a coupling constant of 0.6 MHz was computationally predicted does not allow an easy NMR detection of J(Hg–Hg) owing to its symmetry. The polarization of the Hg_2^{2+} fragment due to an unsymmetric environment would likely cause a reduction of the coupling, though.

In the same work [134] the one-bond and the two-bond Hg–Hg coupling constants in Hg_3^{2+} were calculated. ^2J(Hg-Hg) has been estimated to be much larger than the one-bond coupling. Qualitatively, simple Hückel theory as well as relativistic extended Hückel calculations [136] (P. Pyykkö, unpublished results for Hg_3^{2+}) also suggest that $^2J(\text{Hg–Hg}) > {}^1J(\text{Hg–Hg})$. Again, environmental effects, here from the solvent SO_2 included in the DFT calculations, were found to be very pronounced, thus the ratio $^2J/^1J$ could not yet be predicted with confidence. In such and probably many other metal–metal bonded systems relativistic effects on spin-spin coupling constants serve as a strong "magnifying glass" for the study of the influence of coordination and solvation on the metal–metal bond.

4.5 Miscellaneous Applications: Nuclear-Independent Chemical Shifts and Spin–Spin Coupling Pathways

The expression for the nuclear shielding tensor σ_A does not depend on any particular value of μ_A. Though the spin magnetic moment of the nucleus must not be zero in order for magnetic resonance to occur at all in an experiment, in the theoretical formalism the nuclear shielding can be calculated anyway. In consequence, an explicit expression for σ_A depends only on the position of the nucleus in the molecule. A quantity which has the same formula as the shielding constant can thus be calculated anywhere in space and has been named nuclear independent chemical shift (NICS). The NICS has been extensively applied to the question whether a given system is aromatic. The reader is referred to Volume 101 of Chemical Review for an exhaustive account of aromaticity, in particular regarding various aromaticity criteria [137, 138] and the concept of "ring currents" [139]. Applications to organometallic systems have been carried out. For example, it is known from experiment that the proton chemical shifts in η^6-$(C_6H_6)Cr(CO)_3$ (complex VI) are about 2 ppm smaller than for benzene, raising the question whether the benzene moiety in the complex is less aromatic. This idea got strong support from a computational study of the magnetic susceptibility exaltation of the complex as well as aromatic stabilization energies based on an increment system [140]. It was shown that both criteria indicated a system which is less aromatic than benzene. However, an experimental NMR study from another group had earlier concluded that the complexation of benzene by $Cr(CO)_3$ increases its aromaticity [141]. The supposedly conclusive answer has finally been given by Schleyer et al. [142], who calculated NICS, aromatic stabilization energy and magnetic susceptibility exaltation values for VI and related complexes and found that the NICS criterion for VI indicated a strongly aromatic system. A reanalysis of the procedure in Ref. [140] was further able to reconcile the various criteria for VI. It was concluded that the aromaticity of the benzene moiety in the Cr complex is similar to that of free benzene.

Concepts of a type that could be named nuclear independent spin–spin coupling have to the best of the author's knowledge not yet been proposed. However, recently a number of publications have appeared where "nuclear spin–spin coupling densities" have been employed, for instance, to study pathways of transfer of spin polarization in small molecules [143–145]. It can be expected that such tools will soon be applied to analyze spin-spin coupling tensors in transition-metal complexes.

5 Summary

The calculation of NMR parameters consistently requires the best computational model that is available. However, compromises have to be made regarding the computational efficiency of various methodologies, in order to make the calculations affordable. For transition-metal complexes, the current method of choice is DFT, if necessary including relativistic effects. A number of program codes are available to carry out such computations in an efficient manner, which has led to a large body of available computational data and analyses and interpretations based on these.

Acknowledgments The author is indebted to M. Bühl, M. Kaupp, P. Pyykkö and G. Schreckenbach, who have provided him with unpublished data, preprints of upcoming articles and electronic files for some of the figures.

References

1. Kaupp M, Malkin VG, Malkina OL(1998) In: Schleyer, PvR (ed.) Encyclopedia of computational chemistry. Wiley, Chichester, pp 1857–1866
2. Schreckenbach G, Ziegler, T (1998) Theor Chem Acc 99:71–82
3. Bühl M, Kaupp M, Malkina OL, Malkin VG (1999) J Comput Chem 20:91–105
4. Autschbach J, Ziegler T (2003) Coord Chem Rev 238/239:83–126
5. Bühl M In: Kaupp M, Bühl M, Malkin VG (eds.) Calculation of NMR and EPR parameters. Theory and applications. Wiley-VCH, Weinheim, 2004, pp 421–431
6. Autschbach J, Ziegler T (2002) In: Grant DM, Harris RK (eds.) Encyclopedia of nuclear magnetic resonance, vol 9. Wiley, Chichester, pp 306–323
7. Autschbach, J In: Kaupp M, Bühl M, Malkin VG (eds.) Calculation of NMR and EPR parameters. Theory and applications. Wiley-VCH, Weinheim, 2004, pp 227–247
8. Autschbach J, Ziegler T In: Kaupp M, Bühl M, Malkin VG (eds.) Calculation of NMR and EPR parameters. Theory and applications. Wiley-VCH, Weinheim, 2004, pp 249–264
9. Kaupp M In: Schwerdtfeger P (ed.) Relativistic electronic structure theory, vol 2. Elsevier, Amsterdam (in press)
10. Moon S, Patchkovskii S In: Kaupp M, Bühl M, Malkin VG (eds.) Calculation of NMR and EPR parameters. Theory and applications. Wiley-VCH, Weinheim, 2004, pp 325–351
11. Ruud K, Frediani L, Cammi R, Mennucci B (2003) Int J Mol Sci 4:119–134
12. Bühl M (2002) J Phys Chem A 106:10505–10509
13. Bühl M, Mauschick FT (2002) Phys Chem Chem Phys 4:5508–5514
14. Bühl M, Mauschick FT, Terstegen F, Wrackmeyer B (2002) Angew Chem Int Ed Engl 41:2312–2315
15. Wasylishen R (1996) In: Grant DM, Harris RK (eds.) Encyclopedia of nuclear magnetic resonance. Wiley, pp 1685–1695
16. Wasylishen R In: Kaupp M, Bühl M, Malkin VG (eds.) Calculation of NMR and EPR parameters. Theory and applications. Wiley-VCH, Weinheim, 2004, pp 433–447
17. Gauss J (2000) In: Grotendorst J (ed.) Modern methods and algorithms of quantum chemistry. NIC series, vol 3. John von Neumann Institute for Computing, Jülich, pp 541–592
18. Epstein ST (1974) The variation method in quantum chemistry. Academic, New York
19. Helgaker T, Jaszuński M, Ruud K (1999) Chem Rev 99:293–352
20. Helgaker T, Pecul M In: Kaupp M, Bühl M, Malkin VG (eds.) Calculation of NMR and EPR parameters. Theory and applications. Wiley-VCH, Weinheim, 2004, pp 101–121
21. Fleischer U, van Wüllen C, Kutzelnigg W (1998) In: von Ragué Schleyer P (ed.) Encyclopedia of computational chemistry. Wiley, Chichester, pp 1827–1835
22. McWeeny R (1992) Methods of molecular quantum mechanics. 2nd edn Academic, London
23. Hellmann H (1933) Z Phys 85:180–190
24. Nakatsuji H, Sugimoto M, Saito S (1990) Inorg Chem 29:3095–3097
25. Lee AM, Handy NC, Colwell SM (1995) J Chem Phys 103:10095–10109

26. Malkin VG, Malkina OL, Salahub DR (1993) Chem Phys Letters 204:87–95
27. Malkin VG, Malkina OL, Eriksson LA, Salahub DR (1995) In: Seminario JM, Politzer P (eds.) Modern density functional theory: a tool for chemistry, vol 2. Elsevier, pp 273–347
28. Fadda E, Casida ME, Salahub DR (2003) Int J Quantum Chem 91:67–83
29. Fadda E, Casida ME, Salahub DR (2003) J Chem Phys 118:6758–6768
30. Patchkovskii S, Autschbach J, Ziegler T (2001) J Chem Phys 115:26–42
31. Patchkovskii S, Ziegler T (2002) J Phys Chem A 106:1088–1099
32. Poater J, van Lenthe E, Baerends EJ (2003) J Chem Phys 118:8584–8593
33. Nomura Y, Takeuchi Y, Nakagawa N (1969) Tetrahedron Lett 8:639–642
34. Morishima I, Endo K, Yonezawa T (1973) J Chem Phys 59:3356–3364
35. Kaupp M, Malkina OL, Malkin VG, Pyykkö P (1998) Chem Eur J 4:118–126
36. Almlöf J, Gropen O (1996) In: Lipkowitz KB, Boyd DB (eds.) Reviews in computational chemistry, vol 8. VCH, New York, pp 203–244
37. Pyykkö P (1988) Chem Rev 88:563–594
38. Moss RE (1973) Advanced molecular quantum mechanics. Chapman and Hall, London
39. Sadlej AJ (1994) In: Roos BO (ed.) Lecture notes in chemistry II. Lecture notes in chemistry, vol 64. Springer, Berlin Heidelberg New York, pp 203–230
40. Hess, BA (1998) In: von Ragué Schleyer P (ed.) Encyclopedia of computational chemistry. Wiley, Chichester, pp 2499–2508
41. van Lenthe E, Baerends EJ, Snijders JG (1993) RJ Chem Phys 99:4597–4610
42. van Lenthe E (1996) The ZORA equation. Ph D thesis. Vrije Universiteit Amsterdam, Netherlands
43. Wolff SK, Ziegler T, van Lenthe E, Baerends EJ (1999) J Chem Phys 110:7689–7698
44. Bouten R, Baerends EJ, van Lenthe E, Visscher L, Schreckenbach G, Ziegler, T (2000) J Chem Phys 104:5600–5611
45. Autschbach J, Ziegler T (2000) J Chem Phys 113:936–947
46. Autschbach J, Ziegler T (2000) J Chem Phys 113:9410–9418
47. Fukuda R, Hada M, Nakatsuji H (2003) J Chem Phys 118:1015–1026
48. Fukuda R, Hada M, Nakatsuji H (2003) J Chem Phys 118:1027–1035
49. Fukui H, Baba T (2002) J Chem Phys 117:7836–7844
50. Baba T, Fukui H (2002) Mol Phys 100:623–633
51. Schreckenbach G, Ziegler T (1997) Int J Quantum Chem 61:899–918
52. Wolff SK, Ziegler T (1998) J Chem Phys 109:895–905
53. Khandogin J, Ziegler TA (2000) J Phys Chem A 104:113–120
54. Schwerdtfeger P, Brown JR, Laerdahl JK, Stoll H (2000) J Chem Phys 113:7110–7118
55. Pyykkö P In: Kaupp M, Bühl M, Malkin VG (eds.) Calculation of NMR and EPR parameters. Theory and applications. Wiley-VCH, Weinheim, 2004, pp 7–19
56. Ballhausen, CJ (1979) Molecular electronic structures of transition metal complexes. McGraw-Hill, London
57. De Brouckere G (1978) Adv Chem Phys 37:203–304
58. Tossell JA (ed.) Nuclear magnetic shieldings and molecular structure. Kluwer, Dordrecht
59. Bieger W, Seifert G, Eschrig H, Grossmann G (1985) Chem Phys Letters 115:275–280
60. Freier DG, Fenske RF, Xiao-Zheng Y (1985) J Chem Phys 83:3526–3537
61. Malkin VG, Zhidomirov GM (1988) Zh Strukt Khim 29:32–36
62. Kaupp M, Malkina OL, Malkin VG (1997) J Chem Phys 106:9201–9212
63. Malkin VG, Malkina OL, Salahub DR (1993) Chem Phys Letters 204:80–86

64. Arduengo AJ, Dixon DA, Kumashiro KK, Lee C, Power WP, Zilm KW (1994) J Am Chem Soc 116:6361–6367
65. Schreckenbach G, Ziegler T (1995) J Phys Chem 99:606–611
66. Rauhut G, Puyear S, Wolinski K, Pulay P (1996) J Phys Chem 100:6310–6316
67. Cheeseman JR, Trucks GW, Keith TA, Frisch MJ (1996) J Chem Phys 104:5497–5509
68. Malkin VG, Malkina OL, Salahub DR (1994) Chem Phys Letters 221:91–99
69. Dickson RM, Ziegler T (1996) J Phys Chem 100:5286–5290
70. Helgaker T, Watson M, Handy NC (2000) J Chem Phys 113:9402–9409
71. Sychrovský V, Gräfenstein J, Cremer D (2000) J Chem Phys 113:3530–3547
72. Breit, G (1930) Phys Rev 35:1447–1451
73. (a) Kaupp M, Malkin VG, Malkina OL, Salahub DR (1995) J Am Chem Soc 117:1851–1852; (b) Erratum (1995) J Am Chem Soc 117:8492
74. Kaupp M, Malkin VG, Malkina OL, Salahub DR (1995) Chem Phys Letters 235:382–388
75. Visscher L, Enevoldsen T, Saue T, Jensen HJA, Oddershede, J (1999) J Comput Chem 20:1262–1273
76. Enevoldsen T, Visscher L, Saue T, Jensen HJA, Oddershede, J (2000) J Chem Phys 112:3493–3498
77. Malkin VG, Malkina OL, Salahub DR (1996) Chem Phys Letters 261:335–345
78. Malkina OL, Schimmelpfennig B, Kaupp M, Hess BA, Chandra P, Wahlgren U, Malkin VG (1998) Chem Phys Letters 296:93–104
79. Vaara J, Malkina OL, Stoll H, Malkin VG, Kaupp M (2001) J Chem Phys 114:61–71
80. Koch W, Holthausen MC (2001) A chemist's guide to density functional theory. Wiley-VCH, Weinheim
81. Ruiz-Morales Y, Schreckenbach G, Ziegler T (1996) Organometallics 15:3920–3923
82. Ehlers AW, Ruiz-Morales Y, Baerends EJ, Ziegler T (1997) Inorg Chem 36:5031–5036
83. Ruiz-Morales Y, Schreckenbach G, Ziegler T (1996) J Phys Chem 100:3359–3367
84. Lammertsma K (2003) Top Curr Chem 229:95–119
85. Termaten AJ, Aktas H, Schakel M, Ehlers AW, Lutz M, Spek AL, Lammertsma K (2003) Organometallics 22:1827–1834
86. Webster CE, Hall MB (2002) Inorg Chim Acta 330:268–282
87. Wagener T, Frenking G (1998) Inorg Chem 37:1805–1811
88. Zurek E, Ziegler T (2002) Organometallics 21:83–92
89. Gracia J, Martín A, Mena M, del Carmen Morales-Varela M, Poblet J-M, Santamaría C (2003) Angew Chem Int Ed 42:927–930
90. Strohmeier M, Ohrendt AM, Facelli JC, Solum MS, Pugmire RJ, Parry RW, Grant DM (1997) J Am Chem Soc 119:7114–7120
91. McMahon MT, de Dios AC, Godbout N, Salzmann R, Laws DD, Le H, Havlin RH, Oldfield E (1998) J Am Chem Soc 120:4784–4797
92. Schreckenbach G, Wolff SK, Ziegler T (1999) In: Facelli JC, de Dios AC (eds.) Modeling NMR chemical shifts. (ACS symposium series 732) American Chemical Society, Boston, MA, pp 101–114
93. Schreckenbach G, Wolff SK, Ziegler T (2000) J Phys Chem A 104:8244–8255
94. Schreckenbach, G (2002) Theor Chem Acc 108:246–253
95. Schreckenbach G (2003) Int J Quantum Chem, submitted for publication
96. Bühl M, Mauschick FT, Schurhammer R (2003) In: Wagner S, Handke W, Bode A, Durst F (eds.) High performance computing in science and engineering. Munich 2002. Springer, Berlin Heidelberg New York, pp 189–199
97. Kaupp, M (1996) Habilitationsschrift, Universität Stuttgart, Germany
98. Lichtenberger DL (2003) Organometallics 22:1599–1602

99. Maliarik M, Berg K, Glaser J, Sandström M, Tóth I (1998) Inorg Chem 37:2910–2919
100. Bacskay GB, Bytheway I, Hush NS (1996) J Am Chem Soc 118:3753–3756
101. Craw JS, Bacskay GB, Hush NS (1994): J Am Chem Soc 116:5937–5948
102. Bühl M (1997) Chem Phys Letters 267:251–257
103. Bühl, M (1999) Chem Eur J 5:3514–3522
104. Nakatsuji H, Hu Z-M, Nakajima T (1997) Chem Phys Letters 275:429–436
105. Bühl M (2002) Theor Chem Acc 107:336–342
106. Chan JCC, Au-Yeung CF (1997) J Phys Chem A 101:3637–3640
107. Godbout N, Oldfield E (1997) J Am Chem Soc 119:8065–8069
108. Bühl M, Hamprecht FA (1998) J Comput Chem 19:113–122
109. Kidambi SS, Ramamoorthy A (2003) Inorg Chem 42:2200–2202
110. Kidambi SS, Lee D-K, Ramamoorthy A (2003) Inorg Chem 42:3142–3151
111. Krauss M, Olsen L, Antony J, Hemmingsen L (2002) J Phys Chem B 106:9446–9453
112. Edlund U, Lejon T, Pyykkö P, Venkatachalam TK, Buncel E (1987) J Am Chem Soc 109:5982–5985
113. Rodriguez-Fortea A, Alemany P, Ziegler T (1999) J Phys Chem A 103:8288–8294
114. Hada M, Kaneko H, Nakatsuji H (1996) Chem Phys Letters 261:7–12
115. Autschbach J, Hess BA, Johansson PA, Neugebauer J, Patzschke M, Pyykkö P, Reiher M, Sundholm D (2004) Phys Chem Chem Phys 2004, Vol. 6, pp 11–22
116. Gilbert TM, Ziegler T (1999) J Phys Chem A 103:7535–7543
117. Bühl M, Parrinello M (2001) Chem Eur J 7:4487–4494
118. Marx D, Hutter J (2000) In: Grotendorst, J (ed.) Modern methods and algorithms of quantum chemistry. NIC series, vol 3. John von Neumann Institute for Computing, Jülich, pp 329–477
119. Khandogin J, Ziegler T (1999) Spectrochim Acta A 55:607–624
120. Bryce DL, Wasylishen RE, Autschbach J, Ziegler T (2002) J Am Chem Soc 124:4894–4900
121. Pyykkö P, Pajanne E, Inokuti M (1973) Int J Quantum Chem 7:785–806
122. Pyykkö, P (2000) Theor Chem Acc 103:214–216
123. Autschbach, J, Ziegler, T (2002) J Chem Phys 116:891–896
124. Autschbach J, Ziegler T (2001) J Am Chem Soc 123:3341–3349
125. Autschbach J, Ziegler T (2001) J Am Chem Soc 123:5320–5324
126. Autschbach J, le Guennic B (2003) J Am Chem Soc 125:13585–13593
127. Klamt A, Schüürmann G (1993) J Chem Soc Perkin Trans 2, pp 799–805
128. Pye CC, Ziegler T (1999) Theor Chem Acc 101:396–408
129. Schipper PRT, Gritsenko OV, van Gisbergen SJA, Baerends EJ (2000) J Chem Phys 112:1344–1352
130. Berg KE, Glaser J, Read MC, Tóth I (1995) J Am Chem Soc 117:7550–7751
131. Autschbach J, Igna CD, Ziegler T (2003) J Am Chem Soc 125:1028–1032
132. Boag NM, Browning J, Crocker C, Goggin PL, Goodfellow RJ, Murray M, Spencer JL (1978) J Chem Res Miniprint 2962–2990
133. Xu Q, Heaton BT, Jacob C, Mogi K, Ichihashi Y, Souma Y, Kanamori K, Eguchi T (2000) J Am Chem Soc 122:6862–6870
134. Autschbach J, Igna CD, Ziegler T (2003) J Am Chem Soc 125:4937–4942
135. Malleier R, Kopacka H, Schuh W, Wurst K, Peringer P (2001) Chem Commun 51–52
136. Contreras RH, Facelli JC (1993) Annu Rep NMR Spectrosc 27:255–356
137. Katritzky AR, Jug K, Oniciu DC (2001) Chem Rev 101:1421–1449
138. De Proft F, Geerlings P (2001) Chem Rev 101:1451–1464
139. Gomes JANF, Mallion RB (2001) Chem Rev 101:1349–1383
140. Simion DV, Sorensen TS (1996) J Am Chem Soc 118:7345–7352

141. Mitchell RH, Zhou P, Venugopalan S, Dingle TW (1990) J Am Chem Soc 112:7812–7813
142. Schleyer PvR, Kiran B, Simion DV, Sorensen TS (2000) J Am Chem Soc 122:510–513
143. Soncini A, Lazzeretti P (2003) J Chem Phys 119:1343–1349
144. Malkina OL, Malkin VG (2003) Angew Chem Int Ed 42:4335–4338
145. Wu A, Graefenstein J, Cremer D (2003) J Phys Chem A 107:7043–7056

Structure and Bonding (2004) 112:49–116
DOI 10.1007/b97937

Excitation Energies of Metal Complexes with Time-dependent Density Functional Theory

Angela Rosa[1] · Giampaolo Ricciardi[1] · Oleg Gritsenko[2] · Evert Jan Baerends[2] (✉)

[1] Dipartimento di Chimica, Università della Basilicata, Via N. Sauro 85, 85100 Potenza, Italy
[2] Afdeling Theoretische Chemie, Vrije Universiteit, De Boelelaan 1083, 1081 HV, Amsterdam, The Netherlands
baerends@chem.vu.nl

Abstract After a brief introduction to the theory of time-dependent density functional theory (TDDFT), an overview is given of the recent literature on calculations of the electronic absorption spectra of transition-metal complexes. Various types of complexes are treated – organometallic complexes with carbonyl and cyclopentadienyl ligands, oxo and halide complexes (Werner complexes), metal α-diimine complexes and metallotetrapyrroles. The performance of the various GGA and hybrid functionals is discussed, as well as that of special shape-corrected functionals for the Kohn-Sham potential. The TM complexes are outside of the range of molecules for which B3LYP is parameterized, and this functional does not perform better than pure GGAs. The statistical average of the orbital dependent potentials (SAOP) functional for the Kohn-Sham potential appears to perform best. The strong dependence of the excitation energies on metal-ligand distances is stressed. A consequence is that a full assessment of the accuracy of the TDDFT method will have to include a full vibronic interaction treatment. A conservative estimate is that the accuracy of the best TDDFT methods for TM complexes, where errors of several tenths of an eV often occur, is somewhat lower than for simple benchmark molecules like N_2, CO, and CH_2O, where errors are often of the order of 0.1–0.2 eV.

Keywords TDDFT · Excitation energies · Excited states · Transition metal complexes · Electronic spectra · Metallotetrapyrroles · Metallocarbonyls · α-diimine complexes · Porphyrins · Porphyrazines · Phthalocyanines · Werner complexes · Sandwich complexes

1 Introduction

Electronic spectroscopy of transition metal complexes has been one of the prime experimental methods used to gain understanding about the electronic structure of these systems. The combination of group theory with elementary considerations of the perturbation of atomic orbitals by an external field due to surrounding ligands (crystal field theory) and subsequent elementary orbital considerations (ligand field theory, qualitative molecular orbital theory) has provided a beautiful qualitative framework for rationalizing most spectroscopic features. This theoretical edifice was basically completed some 40 years ago. Successful quantitative prediction of electronic absorption spectra for transition metal complexes has taken much longer to materialize. Major strides towards that goal were actually still being taken in the 1990s. A crucial role has been played by multiconfiguration second-order perturbation theory (CASPT2), which has provided reasonably consistent and accurate results for this type of compound. We refer to the recent review of Pierloot [1] devoted specifically to the application of the CASPT2 method, and the review of Daniel [2] giving an overview over the application of various quantum chemical methods to the excitation spectra of complexes. The present review in a sense complements these other reviews; it focuses on the time-dependent density functional method (TDDFT) that has been implemented for routine application to (large) molecules from 1995 onwards.

Prior to the advent of TDDFT, excitation energies were evaluated in DFT using the ΔSCF method, as proposed in [3]. This method restricts itself to the calculation of energies of single Slater determinants, using available (ground state) functionals. An excited state energy is then obtained in the same way as had been detailed [4, 5] for the restricted Hartree-Fock method, when only single determinantal energies are evaluated (the "sum method"). This approach is not justified by the Hohenberg-Kohn theorem, since it only holds for ground state densities. Nevertheless, a major part of the energy rise of the excited determinant compared to the ground state determinant comes from energy terms such as the kinetic energy and the potential energies for electron-electron and electron-nucleus Coulomb interaction. Those do not depend on a Hohenberg-Kohn theorem. The assumption is then that the exchange-correlation energy of the excited state can be reasonably well represented by a ground state xc energy functional applied to the approximation of the excited state density provided by the excited determinant. One might expect that the Kohn-Sham ΔSCF method would be similar in accuracy to restricted Hartree-Fock ΔSCF calculations. It turns out to be significantly better in most cases. However, a major drawback of this method is that it is essentially lacking any configuration mixing. In excited states that have an essential mixing of configurations, the ΔSCF method must fail, if not for the numerical result for the excitation energy, then surely for the character of the excited state.

The TDDFT method is a significant advance over the ΔSCF method in two respects. First, it has a sound theoretical basis in the extension of the Hohenberg-Kohn theorem to systems in a time-varying field [6]. Second, it can be implemented with a full account of symmetry aspects, so as to automatically yield a given number of lowest excited states belonging to a certain irreducible representation of the molecular point group, rendering excitation energy calculations extremely simple and straightforward to execute. Third, the TDDFT method is also more accurate than the ΔSCF method, although this is not universally so, and in particular for $d \rightarrow d$ transitions, for which the ΔSCF method is often quite accurate, the superiority of the TDDFT method has not been unequivocally established.

In this paper, we will review the application of TDDFT to a number of prototype TM complexes. First, in Sect. 2, the theory behind the TDDFT approach will be summarized. From the basic equations, it will be apparent that the orbital energies of the KS orbitals play a crucial role, as well as the MO shapes themselves. It will be seen that it is therefore important to distinguish between exact Kohn-Sham DFT results, and approximations such as the local-density approximation (LDA) and the generally applied generalized gradient approximations (GGA). It is crucial that the ground state Kohn-Sham MO calculations, which are the basis of the TDDFT method, are more accurate than is attainable with the LDA and GGA methods. This is related to the fact that the LDA and GGA Kohn-Sham potentials have two important deficiencies. First, they are too shallow in the molecular region, so the orbital energies of the occupied orbitals, and of the lowest unoccupied orbitals if they have their amplitude in the molecular domain, are shifted upwards by ~5 eV. The orbital energies of unoccupied levels close to zero energy, such as the Rydberg levels, cannot shift up much. Excitations to those Rydberg levels will therefore typically be (much) too low in LDA or GGA calculations. Second, the asymptotic decay to zero of the KS potentials is not Coulombic ($-1/r$) for LDA and GGA. The much too fast (exponential) decay of these approximate potentials distorts the shape of diffuse unoccupied orbitals, in particular Rydberg orbitals, and therefore yields errors in intensities. We will discuss the attempts that have been made to improve upon the LDA and GGA approximations for TDDFT excitation energy calculations.

In Sects. 3–6, various types of TM complexes will be discussed. We start in Sect. 3 with elementary organometallic complexes – the metallocarbonyls and some cyclopentadienyl complexes. In Sect. 4 classical Werner complexes are dealt with. A separate section (Sect. 5) is devoted to the important class of technologically important and widely studied polypyridyl complexes. Section 6 deals with metallotetrapyrroles such as porphyrins and phthalocyanines. Section 7 contains concluding remarks.

2 Theory

2.1 Basic Equations of TDDF(R)T

The extension of density functional theory (DFT) to the dynamical description of atomic and molecular systems offers an efficient theoretical and computational tool for chemistry and molecular spectroscopy, namely, time-dependent DFT (TDDFT) [7–11]. This tool allows us to simulate the time evolution of electronic systems, so that changes in molecular structure and bonding over time due to applied time-dependent fields can be investigated. Its response variant TDDF(R)T is used to calculate frequency-dependent molecular response properties, such as polarizabilities and hyperpolarizabilities [12–17]. Furthermore, TDDFRT overcomes the well-known difficulties in applying DFT to excited states [18], in the sense that the most important characteristics of excited states, the excitation energies and oscillator strengths, are calculated with TDDFRT [17, 19–26].

Until recently, the development and applications of TDDFT were directed, chiefly, at atoms and prototype molecules of elements from main groups of the periodic table. Application of TDDFT to transition metal complexes remains a challenge, and this review article presents recent examples of such applications. All chemical applications share the same basic theory and approximations, the main features of which will be outlined below.

Time-dependent density functional methods were applied in the early 1980s by Zangwill and Soven [27], Deb and Ghosh [28], Stott and Zaremba [29], Mahan [30], and Bartolotti and Achariya [31]. The same theoretical basis as time-independent DFT was provided in 1984 by the theorem of Runge and Gross [6], which considers the time evolution of the electron density $\rho(\mathbf{r}_1,t)$ of a many-electron system represented at the initial time t=0 by a definite wavefunction $\Psi(\mathbf{x}_1,...,\mathbf{x}_N,t=0)$ ($\mathbf{x}_i=[\mathbf{r}_i,s_i]$, where $\mathbf{r}_i$ are the spatial and s_i are the spin electron coordinates). The density $\rho(\mathbf{r}_1,t)$ evolves under the influence of the external potential $v_{ext}(\mathbf{r}_1,t)$. The Runge-Gross theorem establishes, that, given the wavefunction at t=0, at all later times t>0 the density $\rho(\mathbf{r}_1,t)$ determines the potential $v_{ext}(\mathbf{r}_1,t)$ uniquely up to an additive purely time-dependent function. In turn, the potential $v_{ext}(\mathbf{r}_1,t)$ uniquely determines the wavefunction $\Psi(\mathbf{x}_1,...,\mathbf{x}_N,t)$ of the evolving system, which can therefore be considered to be a functional of the time-dependent density, $\Psi(\mathbf{x}_1,...,\mathbf{x}_N,t)=\Psi([\rho],t)$. In other words, a unique mapping

$$\rho(\mathbf{r}_1,t) \Leftrightarrow v_{ext}(\mathbf{r}_1,t) \Leftrightarrow \Psi(\mathbf{x}_1,t,...\mathbf{x}_N,t) \tag{1}$$

is established between $\rho(\mathbf{r}_1,t)$, $v_{ext}(\mathbf{r}_1,t)$, and $\Psi(\mathbf{x}_1,...,\mathbf{x}_N,t)$.

In the leading Kohn-Sham (KS) approach of TDDFT [7] the density $\rho(\rho = \rho_\alpha + \rho_\beta, \rho_\sigma$ is the spin-density) is represented in the orbital form

$$\rho_\sigma(\mathbf{r}_1,t) = \sum_i^{N_\sigma} |\phi_{i\sigma}(\mathbf{r}_1,t)|^2 \tag{2}$$

through the sum of the densities of the KS spin-orbitals $\phi_{i\sigma}$, the first N_σ solutions of the time-dependent one-electron KS equations (atomic units are used)

$$\left\{-\frac{1}{2}\nabla^2 + v_{s\sigma}(\mathbf{r}_1,t)\right\}\phi_{i\sigma}(\mathbf{r}_1,t) = i\frac{\partial}{\partial t}\phi_{i\sigma}(\mathbf{r}_1,t). \tag{3}$$

The orbitals $\phi_{i\sigma}(\mathbf{r}_1,t)$ evolve under the influence of the KS potential $v_{s\sigma}$, which consists of the external potential v_{ext} and the potential $v_{el\sigma}$ of the electron-electron interaction

$$v_{s\sigma}(\mathbf{r}_1,t) = v_{ext}(\mathbf{r}_1,t) + v_{el\sigma}(\mathbf{r}_1,t). \tag{4}$$

In turn, $v_{el\sigma}$ is subdivided into the standard Hartree potential v_{Coul} of electrostatic electron repulsion and the exchange-correlation (xc) potential $v_{xc\sigma}$

$$v_{el\sigma}(\mathbf{r}_1,t) = v_{Coul}(\mathbf{r}_1,t) + v_{xc\sigma}(\mathbf{r}_1,t). \tag{5}$$

The important advantage of the KS approach is its efficient treatment of the electron correlation, in the sense that all effects of the electron exchange and Coulomb correlation are incorporated into the local (in other words the same for all orbitals and multiplicative) xc potential $v_{xc\sigma}$.

The time-dependent KS Eq. 3 can be used to simulate the full time evolution of the density. However, in this review article we shall exclusively consider the linear response TDDF(R)T approach. Its starting point is Eq. 3 with the time-independent potential $v_{s\sigma}(\mathbf{r}_1)$. In this case, the orbitals $\phi_{i\sigma}(\mathbf{r}_1,t)$ only have a phase factor that oscillates in time

$$\phi_{i\sigma}(\mathbf{r}_1,t) = \phi_{i\sigma}(\mathbf{r}_1)e^{-i\varepsilon_{i\sigma}t}. \tag{6}$$

Their spatial parts $\phi_{i\sigma}(\mathbf{r}_1)$ and orbital energies $\varepsilon_{i\sigma}$ are defined from the equations for the stationary ground-state KS system

$$\left\{-\frac{1}{2}\nabla^2 + v_{\text{ext}}(\mathbf{r}_1) + v_{\text{Coul}}(\mathbf{r}_1) + v_{\text{xc}\sigma}(\mathbf{r}_1)\right\}\phi_{i\sigma}(\mathbf{r}_1) = \varepsilon_{i\sigma}\phi_{i\sigma}(\mathbf{r}_1). \tag{7}$$

TDDFRT considers a linear response of the ground-state density to the time-dependent perturbation of the external potential $\delta v_{\text{ext}}(\mathbf{r}_1,t)$ which is, usually, a harmonic perturbation $\delta v_{\text{ext}}(\mathbf{r}_1,t)$ switched on at time t_0

$$\delta v_{\text{ext}}(\mathbf{r}_1,t) = \delta v_{\text{ext}}(\mathbf{r}_1)\cos(\omega t)\Theta(t-t_0) \tag{8}$$

Due to this perturbation, the electron density will acquire a time-dependent change. In the same way as for the time-independent system we assume that an equivalent system of noninteracting electrons moving in a local potential (now time-dependent) can be found, the orbital densities of which build the exact total time-dependent density. We write the KS potential as the sum of the original KS potential plus the time-dependent perturbation, $v_{s\sigma}(\mathbf{r}_1) + \delta v_{s\sigma}(\mathbf{r}_1,t)$ and the KS orbitals as $\phi_{i\sigma}(\mathbf{r}_1)\exp(-i\varepsilon_{i\sigma}t) + \delta\phi_{i\sigma}(\mathbf{r}_1,t)$. The time-dependent KS system satisfies the time-dependent one-electron Schrödinger equation

$$\left\{-\frac{1}{2}\nabla^2 + v_{s\sigma}(\mathbf{r}_1) + \delta v_{s\sigma}(\mathbf{r}_1,t) - i\frac{\partial}{\partial t}\right\}\left[\phi_{i\sigma}(\mathbf{r}_1)e^{-i\varepsilon_{i\sigma}t} + \delta\phi_{i\sigma}(\mathbf{r}_1,t)\right] = 0 \tag{9}$$

Following the standard treatment of perturbation theory, the orbital change is expanded in terms of the unperturbed orbitals (Eq. 6)

$$\delta\phi_{i\sigma}(\mathbf{r}_1,t) = \sum_{j\neq i} c_{ij}^{\sigma}(t)\phi_{j\sigma}(\mathbf{r}_1)e^{-i\varepsilon_{j\sigma}t}. \tag{10}$$

where we only consider such weak perturbations or such short times that the coefficients $c_{ij\sigma}(t)$ remain small first order quantities (the linear response regime). Insertion of (10) into (9), and simplifying through the application of Eq. 3, leads to the set of equations

$$\begin{aligned}\sum_{j\neq i}\frac{dc_{ij}^{\sigma}(t)}{dt}\phi_{j\sigma}(\mathbf{r}_1)e^{-i\varepsilon_{j\sigma}t} &= -i\delta v_{s\sigma}(\mathbf{r}_1,t)\phi_{i\sigma}(\mathbf{r}_1,t)e^{-i\varepsilon_{i\sigma}t}\\ &\quad -i\sum_{j\neq i} c_{ij}^{\sigma}(t)\delta v_{s\sigma}(\mathbf{r}_1,t)\phi_{j\sigma}(\mathbf{r}_1)e^{-i\varepsilon_{j\sigma}t}\end{aligned} \tag{a}$$

Multiplication by $\phi_{k\sigma}^{*}(\mathbf{r}_1)$ and integration over $\mathbf{r}_1$ yields the following expression for the expansion coefficients $c_{ij\sigma}(t)$, where the coupled equations for the coefficients $c_{ij\sigma}(t)$ are decoupled by neglecting the last term on the right hand side of the above equation, on account of the assumption that $c_{ij\sigma}(t)$ are only first order quantities

$$c_{ij}^{\sigma}(t) = -i \int_{-\infty}^{t} dt' \int d\mathbf{r}_1 \phi_{j\sigma}^{*}(\mathbf{r}_1) \delta v_{s\sigma}(\mathbf{r}_1, t') \phi_{i\sigma}(\mathbf{r}_1) e^{i(\varepsilon_{j\sigma} - \varepsilon_{i\sigma})t'}. \tag{11}$$

From (10) and (11), the density response is obtained as follows

$$\begin{aligned}
\delta\rho_{\sigma}(\mathbf{r}_1, t) &= \sum_{i}^{N_\sigma} \left\{ \phi_{i\sigma}^{*}(\mathbf{r}_1) e^{i\varepsilon_{i\sigma}t} \delta\phi_{i\sigma}(\mathbf{r}_1, t) + \delta\phi_{i\sigma}^{*}(\mathbf{r}_1, t) \phi_{i\sigma}(\mathbf{r}_1) e^{-i\varepsilon_{i\sigma}t} \right\} \\
&= -i \int d\mathbf{r}_2 \int_{-\infty}^{t} dt' \delta v_{s\sigma}(\mathbf{r}_2, t') \sum_{i}^{N_\sigma} \sum_{j \neq i} \left\{ \phi_{i\sigma}^{*}(\mathbf{r}_1) \phi_{j\sigma}(\mathbf{r}_1) \phi_{j\sigma}^{*}(\mathbf{r}_2) \phi_{i\sigma}(\mathbf{r}_2) \cdot \right. \\
&\quad \left. \cdot e^{-i(\varepsilon_{j\sigma} - \varepsilon_{i\sigma})(t - t')} . - \phi_{i\sigma}(\mathbf{r}_1) \phi_{j\sigma}^{*}(\mathbf{r}_1) \phi_{j\sigma}(\mathbf{r}_2) \phi_{i\sigma}^{*}(\mathbf{r}_2) e^{-i(\varepsilon_{i\sigma} - \varepsilon_{j\sigma})(t - t')} \right\}
\end{aligned} \tag{12}$$

Usually, the frequency-dependent response properties are of interest, so that the frequency-dependent density response $\delta\rho_{\sigma}(\mathbf{r}_1, \omega)$ to the potential change $\delta v_{s\sigma}(\mathbf{r}_2, \omega)$ is considered in TDDFRT. It is obtained with the Fourier transformation of $\delta\rho_{\sigma}(\mathbf{r}_1, t)$ of (12)

$$\delta\rho_{\sigma}(\mathbf{r}_1, \omega) = \int_{-\infty}^{\infty} \delta\rho_{\sigma}(\mathbf{r}_1, t)\, e^{-i\omega t} dt \tag{13}$$

In order to carry out the transformation, one can take the following auxiliary integral, which accumulates the time-dependent terms with $p=t-t'$ of (12)

$$\lim_{\eta \to 0} \int_{-\infty}^{\infty} \Theta(p) e^{-i(\varepsilon_{j\sigma} - \varepsilon_{i\sigma})p} e^{i\omega p} e^{-\eta p} dp = \lim_{\eta \to 0} \frac{1}{\omega + i\eta - (\varepsilon_{j\sigma} - \varepsilon_{i\sigma})} \tag{14}$$

In (14) the Heavyside function $\Theta(p)$, $\Theta(t-t')=1$ for $t' \le t$ and $\Theta(t-t')=0$ for $t'>t$, incorporates the limits of the integral over the time in (11). Also, a factor $\exp(-\eta p)$ is introduced in the integrand of (14) in order to make the integral convergent and the limit $\eta \to 0$ is taken.

With (14), the Fourier transform of Eq. 12 turns to a key expression of TDDFRT for the density response $\delta\rho_{\sigma}(\mathbf{r}_1, \omega)$

$$\delta\rho_{\sigma}(\mathbf{r}_1, \omega) = \int \chi_{\sigma}^{s}(\mathbf{r}_1, \mathbf{r}_2, \omega) \delta v_{s\sigma}(\mathbf{r}_2, \omega) d\mathbf{r}_2, \tag{15}$$

where χ_{σ}^{s} is a simplified linear response function, connecting the response of the diagonal density to a perturbation by a local multiplicative potential

$$\chi_\sigma^s(\mathbf{r}_1,\mathbf{r}_2,\omega)=\lim_{\eta\to 0}\sum_i^{N_\sigma}\sum_{j\neq i}\left\{\frac{\phi_{i\sigma}^*(\mathbf{r}_1)\phi_{j\sigma}(\mathbf{r}_1)\phi_{j\sigma}^*(\mathbf{r}_2)\phi_{i\sigma}(\mathbf{r}_2)}{\omega+i\eta-\left(\varepsilon_{j\sigma}-\varepsilon_{i\sigma}\right)}\right.$$
$$\left.-\frac{\phi_{i\sigma}(\mathbf{r}_1)\phi_{j\sigma}^*(\mathbf{r}_1)\phi_{j\sigma}(\mathbf{r}_2)\phi_{i\sigma}^*(\mathbf{r}_2)}{\omega+i\eta+\left(\varepsilon_{j\sigma}-\varepsilon_{i\sigma}\right)}\right\}$$
$$=\lim_{\eta\to 0}\sum_i\sum_j\left(n_{i\sigma}-n_{j\sigma}\right)\frac{\phi_{i\sigma}^*(\mathbf{r}_1)\phi_{j\sigma}(\mathbf{r}_1)\phi_{j\sigma}^*(\mathbf{r}_2)\phi_{i\sigma}(\mathbf{r}_2)}{\omega+i\eta-\left(\varepsilon_{j\sigma}-\varepsilon_{i\sigma}\right)}. \tag{16}$$

In the last line of (16) $n_{i\sigma}$ is the occupation of the orbital $\phi_{k\sigma}$. Inserting (16) into (15) and defining δP_{ij}^σ

$$\delta P_{ij}^\sigma(\omega)=\chi_{ij\sigma}^s(\omega)\delta v_{s\sigma}^{ji}(\omega) \tag{17}$$

where $\chi_{ij\sigma}^s(\omega)$ is the element of the response function

$$\chi_{ij\sigma}^s(\omega)=\frac{n_{i\sigma}-n_{j\sigma}}{\omega-\left(\varepsilon_{j\sigma}-\varepsilon_{i\sigma}\right)} \tag{18}$$

(where the limit $\eta\to 0$ has been taken), we obtain the density change as

$$\delta\rho(\mathbf{r}_1,\omega)=\sum_{i,j}\delta P_{ij}^\sigma(\omega)\phi_{i\sigma}^*(\mathbf{r}_1)\phi_{j\sigma}(\mathbf{r}_1). \tag{b}$$

Due to the initial perturbation (8) of the external potential, another component of $v_{s\sigma}$ of (4), the potential $v_{el\sigma}$ of (5), also acquires a change, so that the matrix element $\delta v_{s\sigma}^{ij}$ in (17) is the following sum

$$\delta v_{s\sigma}^{ij}(\omega)=\delta v_{\text{ext}}^{ij\sigma}(\omega)+\delta v_{\text{el}}^{ij\sigma}(\omega). \tag{19}$$

While the matrix elements of the external potential $\delta v_{\text{ext}}^{ij\sigma}$ are simply

$$\delta v_{\text{ext}}^{ij\sigma}(\omega)=\int\phi_{i\sigma}^*(\mathbf{r}_1)\delta v_{\text{ext}}(\mathbf{r}_1,\omega)\phi_{j\sigma}(\mathbf{r}_1)d\mathbf{r}_1, \tag{20}$$

the matrix elements of the potential of electron-electron interaction $\delta v_{el}^{ij\sigma}$ depend themselves on the density changes δP_{kl}^τ

$$\delta v_{el}^{ij\sigma}(\omega)=\sum_{kl\tau}K_{ij\sigma,kl\tau}(\omega)\delta P_{kl}^\tau(\omega). \tag{21}$$

The connection between $\delta v_{el}^{ij\sigma}$ and δP_{kl}^τ is carried out with the coupling matrix $K_{ij\sigma,kl\tau}$, the latter, just like the potential $v_{el\sigma}$, consists of the Hartree $K_{ij\sigma,kl\tau}^{\text{Coul}}$ and xc $K_{ij\sigma,kl\tau}^{\text{xc}}$ parts [32]

$$K_{ij\sigma,kl\tau}(\omega)=K_{ij\sigma,kl\tau}^{\text{Coul}}+K_{ij\sigma,kl\tau}^{\text{xc}}(\omega) \tag{22}$$

The former part contains the operator of electron-electron interaction $1/r_{12}$

$$K^{\mathrm{Coul}}_{ij\sigma,kl\tau} = \int \phi^*_{i\sigma}(\mathbf{r}_1)\phi_{j\sigma}(\mathbf{r}_1)\frac{1}{r_{12}}\phi_{k\tau}(\mathbf{r}_2)\phi^*_{l\tau}(\mathbf{r}_2)d\mathbf{r}_1 d\mathbf{r}_2, \tag{23}$$

$$K^{xc}_{ij\sigma,kl\tau} = \int \phi^*_{i\sigma}(\mathbf{r}_1)\phi_{j\sigma}(\mathbf{r}_1)f^{\sigma\tau}_{xc}(\mathbf{r}_1,\mathbf{r}_2,\omega)\phi_{k\tau}(\mathbf{r}_2)\phi^*_{l\tau}(\mathbf{r}_2)d\mathbf{r}_1 d\mathbf{r}_2, \tag{24}$$

while the change in the xc potential is determined with the xc kernel $f^{\sigma\tau}_{\mathrm{xc}}$the latter being the functional derivative of the xc potential $v_{\mathrm{xc}\sigma}$with respect to the spin-density ρ_τ

$$f^{\sigma\tau}_{xc}(\mathrm{r}_1,\mathrm{r}_2,t-t') = \frac{\delta v_{xc\sigma}(\mathrm{r}_1,t)}{\delta\rho_\tau(\mathrm{r}_2,t')}. \tag{25}$$

With the dependence (21) of $\delta v^{ij\sigma}_{el}$ on δP^{τ}_{kl}, the evaluation of the density response through (17) becomes a self-consistent procedure.

2.2 Excitation Energies with TDDFRT

Perhaps the most remarkable ability of TDDFRT is its power to calculate atomic and molecular electronic excitations [9]. TDDFRT accomplishes this without direct calculation of excited states, so that their densities are not available from TDDFRT. Instead, harmonic external perturbations are applied to a many-electron system, and from the response of the ground-state density the energies are evaluated in TDDFRT of excitations that *might* occur in the system. Since the theory treats the exact density response, exact excitation energies and corresponding oscillator strengths could be, in principle, determined in TDDFRT. In this sense, TDDFRT circumvents the well-known problem of the proper description of excited states in DFT.

To determine excitation energies, it is sufficient to consider the real KS orbitals within the theory of the preceding section. In the response function (16) we replace the complex orbitals with real ones and use the fact that the terms with the occupied orbitals $\phi_{i\sigma}$ and $\phi_{j\sigma}$ cancel each other in (16). We also drop the term $i\eta$ in the denominator. From this follows the real form of the response function (we use the indices i and j for the occupied orbitals, while the indices a and b are used for the virtual ones)

$$\chi^s_\sigma(\mathbf{r}_1,\mathbf{r}_2,\omega) = \sum_i^{N_\sigma}\sum_a \phi_{i\sigma}(\mathbf{r}_1)\phi_{a\sigma}(\mathbf{r}_1)\phi_{i\sigma}(\mathbf{r}_2)\phi_{a\sigma}(\mathbf{r}_2)\left\{\frac{1}{\varepsilon_{i\sigma}-\varepsilon_{a\sigma}+\omega} + \frac{1}{\varepsilon_{i\sigma}-\varepsilon_{a\sigma}-\omega}\right\} = \sum_i^{N_\sigma}\sum_a \phi_{i\sigma}(\mathbf{r}_1)\phi_{a\sigma}(\mathbf{r}_1)\phi_{i\sigma}(\mathbf{r}_2)\phi_{a\sigma}(\mathbf{r}_2)\frac{2(\varepsilon_{i\sigma}-\varepsilon_{a\sigma})}{(\varepsilon_{i\sigma}-\varepsilon_{a\sigma})^2-\omega^2} \tag{26}$$

Then, the corresponding elements $\chi^s_{ij\sigma}$ in the density response matrix equation (17) become

$$\chi^s_{ia\sigma}(\omega) = \frac{(\varepsilon_{i\sigma} - \varepsilon_{a\sigma})}{(\varepsilon_{i\sigma} - \varepsilon_{a\sigma})^2 - \omega^2} \tag{27}$$

Insertion of (27) as well as the potential changes (19) to (21) into (17) yields a more detailed equation

$$\delta P^\sigma_{ia}(\omega) = \frac{(\varepsilon_{i\sigma} - \varepsilon_{a\sigma})}{(\varepsilon_{i\sigma} - \varepsilon_{a\sigma})^2 - \omega^2} \left[\delta v^{ia\sigma}_{\text{ext}}(\omega) + 2 \sum_{jb\tau} K_{ia\sigma,jb\tau}(\omega) \delta P^\tau_{jb}(\omega) \right] \tag{28}$$

Rearranging (28) and separating its free terms, we arrive at the following matrix equation for the density response

$$\sum_{jb\tau} \left[M_{ia\sigma,jb\tau}(\omega) + \omega^2 S_{ia\sigma,jb\tau} \right] \delta P^\tau_{jb}(\omega) = \delta v^{ia\sigma}_{\text{ext}}(\omega) \tag{29}$$

where $M_{ia\sigma,jb\tau}$ are the following matrix elements

$$M_{ia\sigma,jb\tau}(\omega) = \delta_{\sigma\tau}\delta_{ab}\delta_{ij}(\varepsilon_{j\tau} - \varepsilon_{b\tau}) - 2K_{ia\sigma,jb\tau}(\omega), \tag{30}$$

$$S_{ia\sigma,jb\tau} = \frac{\delta_{\sigma\tau}\delta_{ab}\delta_{ij}}{\varepsilon_{j\tau} - \varepsilon_{b\tau}}. \tag{31}$$

To advance further, the resonance property of the exact density response at the true excitation energies ω_k [9, 20, 32] is used, namely, that a finite external perturbation $\delta v_{ext}(\mathbf{r}_1, \omega_k)$ leads to an infinite change in the density. Then Eq. 29 can be satisfied only if the matrix in the square brackets possesses a zero eigenvalue at ω_k. This leads, after a unitary transformation, to the master matrix eigenvalue equations for the excitation energies ω_k [9]

$$\mathbf{\Omega F}_k = \omega_k^2 \mathbf{F}_k, \tag{32}$$

In (32), the matrix Ω is defined through the matrices (30) and (31) as follows

$$\Omega = -\mathbf{S}^{-1/2}\mathbf{M}\mathbf{S}^{-1/2}, \tag{33}$$

so that its matrix elements are

$$\Omega_{ia\sigma,jb\tau} = \delta_{\sigma\tau}\delta_{ab}\delta_{ij}(\varepsilon_{b\tau} - \varepsilon_{j\tau})^2 + 2\sqrt{\varepsilon_{a\sigma} - \varepsilon_{i\sigma}} K_{ia\sigma,jb\tau} \sqrt{\varepsilon_{b\tau} - \varepsilon_{j\tau}}. \tag{34}$$

From (32) and (34) it follows that the zero-order TDDFRT estimate $\omega_k{}^0$ for the excitation energy ω_k when the coupling matrix is neglected, K=0, is just the energy difference

$$\omega_k^0 = \varepsilon_{a\sigma} - \varepsilon_{i\sigma} \tag{35}$$

between the virtual $\phi_{a\sigma}$ and the occupied $\phi_{i\sigma}$ orbitals involved in the electron excitation. The second term in the r.h.s. of (34) brings an important correction from the electron coupling, which shifts $\omega^0{}_k$ to the true excitation energy ω_k.

The eigenvectors $\mathbf{F}_k$ determine the oscillator strengths of the excitations. This can be established using the sum-over-states representation of the standard many-electron theory for the dynamic dipole polarizability $\alpha_{av}(\omega)$ [33]

$$\alpha_{av}(\omega) = \sum_k \frac{f_k}{\omega_k^2 - \omega^2} \tag{36}$$

where f_k are the spectroscopic oscillator strengths. One can compare (36) with the corresponding TDDFRT formula [9]

$$\alpha_{av}(\omega) = \frac{2}{3} \sum_k \sum_l^3 \frac{\left| \mathbf{r}_l^+ \mathbf{S}^{-1/2} \mathbf{F}_k \right|^2}{\omega_k^2 - \omega^2}, \tag{37}$$

where $\mathbf{r}^+{}_l$ is the adjoint of the matrix

$$r_l^{ij\sigma} = \int \phi_{i\sigma}^*(\mathbf{r}_1) r_l \phi_{j\sigma}(\mathbf{r}_1) d\mathbf{r}_1 \tag{38}$$

with r_l being the Cartesian directions *x,y,z*. Comparison of (37) and (38) allows us to express the oscillator strengths f_k through the eigenvectors $\mathbf{F}_k$ of (32)

$$f_k = \sum_l^3 \left| \mathbf{r}_l^+ \mathbf{S}^{-1/2} \mathbf{F}_k \right|^2. \tag{39}$$

In the important case of a closed-shell system, the matrix Ω of Eqs. 32 to 34 of the spin-restricted TDDFRT can be split into two parts Ω^{S} and Ω^{T} representing the singlet-singlet and singlet-triplet excitations [10, 34]

$$\begin{aligned} \Omega^{\mathrm{S}}_{ia,jb} &= \delta_{ab}\delta_{ij}\left(\varepsilon_b - \varepsilon_j\right)^2 \\ &\quad + 2\sqrt{\varepsilon_a - \varepsilon_i}\left[2K^{\mathrm{Coul}}_{ia,jb} + K^{\mathrm{xc}(\uparrow\uparrow)}_{ia,jb}(\omega) + K^{\mathrm{xc}(\uparrow\downarrow)}_{ia,jb}(\omega)\right]\sqrt{\varepsilon_b - \varepsilon_j} \end{aligned} \tag{40}$$

$$\Omega^{\mathrm{T}}_{ia,jb} = \delta_{ab}\delta_{ij}\left(\varepsilon_b - \varepsilon_j\right)^2 + 2\sqrt{\varepsilon_a - \varepsilon_i}\left[K^{\mathrm{xc}(\uparrow\uparrow)}_{ia,jb}(\omega) - K^{\mathrm{xc}(\uparrow\downarrow)}_{ia,jb}(\omega)\right]\sqrt{\varepsilon_b - \varepsilon_j}. \tag{41}$$

In (40) and (41), the arrows attached to the xc elements of the coupling matrix $K_{ia,jb}{}^{\mathrm{xc}}$ indicate the relative orientation of the spins σ and τ for the corresponding components of the xc kernel.

The TDDFRT equations presented in this section offer an efficient way to calculate the key characteristics of excited states, the excitation energies ω_k, and the corresponding oscillator strengths. In the next sections the applications of this theory will be discussed.

2.3 The xc Potential and Kernel

As was mentioned above, in KS-TDDFT the effects of electron exchange and Coulomb correlation are incorporated in the exchange-correlation potential $v_{xc\sigma}$ and kernel $f_{xc}^{\sigma\tau}$. While the potential determines the KS orbitals $\phi_{i\sigma}$ and the zero-order TDDFRT estimate (35) for excitation energies, the kernel determines the change of $v_{xc\sigma}$ with Eqs. 21, 22, 24. Though both $v_{xc\sigma}$ and $f_{xc}^{\sigma\tau}$ are well defined in the theory, their exact explicit form as functionals of the density is not known. Rather accurate $v_{xc\sigma}$ potentials can be constructed numerically from the ab initio densities ρ for atoms [35–38] and molecules [39–42]. However, this requires tedious correlated ab initio calculations, usually with some type of configuration interaction (CI) method. Therefore, approximations to $v_{xc\sigma}$ and $f_{xc}^{\sigma\tau}$ are to be used in TDDFT.

The standard approximations in TDDFT are the local density approximation (LDA) for $v_{xc\sigma}$ and the related adiabatic LDA (ALDA) for $f_{xc}^{\sigma\tau}$. Within LDA, the potential $v_{xc\sigma}^{\mathrm{LDA}}$ is taken from the model of the homogeneous electron gas [43], with the dominant exchange part $v_{x\sigma}^{\mathrm{LDA}}$ of $v_{xc\sigma}^{\mathrm{LDA}}$ being only the 1/3 power of ρ_σ

$$v_{x\sigma}^{\mathrm{LDA}}([\rho_\sigma];\mathbf{r}_1) = -2\left(\frac{3}{4\pi}\right)^{1/3} \rho_\sigma^{1/3}(\mathbf{r}_1). \tag{42}$$

In turn, the ALDA kernel $f_{xc}^{\sigma\tau(\mathrm{ALDA})}$ is obtained as the time-independent and spatially local derivative of $v_{xc\sigma}^{\mathrm{LDA}}$

$$f_{xc}^{\sigma\tau(ALDA)}(\mathbf{r}_1,\mathbf{r}_2) = \delta(\mathbf{r}_1-\mathbf{r}_2)\left.\frac{dv_{xc\sigma}^{\mathrm{LDA}}}{d\rho_\tau}\right|_{\rho_\tau(\mathbf{r}_1)} \tag{43}$$

with the exchange part

$$f_{x}^{\sigma\sigma(\mathrm{ALDA})}(\mathbf{r}_1,\mathbf{r}_2) = \delta(\mathbf{r}_1-\mathbf{r}_2)\frac{dv_{x\sigma}^{\mathrm{LDA}}}{d\rho_\sigma}\Big|_{\rho_\sigma(\mathbf{r}_1)} = -\frac{2}{3}\left(\frac{3}{4\pi}\right)^{1/3} \rho_\sigma^{-2/3}(\mathbf{r}_1) \tag{44}$$

Though LDA/ALDA calculations with $v_{xc\sigma}^{\mathrm{LDA}}$ and $f_{xc}^{\sigma\tau(\mathrm{ALDA})}$ yield a decent estimate of molecular polarizabilities and electron excitations, higher quality approximations are required for many chemical applications.

The development of TDDFT has shown that, in order to obtain better results, getting closer to the accurate xc potential $v_{xc\sigma}$ is relatively more important than better modeling of the xc kernel $f_{xc}^{\sigma\tau}$. Some benchmark results have been obtained in [24], where an essentially accurate v_{xc} constructed numerically from the ab initio (CI) density ρ for some closed-shell prototype sys-

Table 1 Molecular static isotropic polarizabilities α obtained with the LDA, BP and SAOP potentials and with the accurate v_{xc} (see [17])

Molecule	LDA	BP	SAOP	Accurate v_{xc}	Reference data
H_2	5.90	5.55	5.63	5.16	5.18
N_2	12.27	12.06	11.82	11.68	11.74
HF	6.20	6.10	5.40	5.49	5.52
HCl	18.63	18.11	17.99	17.25	17.39
H_2O	10.53	10.33	9.45	9.45	9.64
CO	13.87	13.38	13.01	12.86	13.08
Mean absolute error	8.8%	5.6%	2.9%	1.0%	–

tems, was combined within TDDFRT with the ALDA kernel (43). Table 1 compares static isotropic polarizabilities α calculated with (accurate v_{xc})/ALDA and LDA/ALDA with the benchmark reference data. Table 1 also presents α calculated with the xc potential v_{xc}^{BP} of a standard generalized gradient approximation (GGA) of DFT (the combination BP of the GGA exchange potential of Becke (B88) [44] and the correlation potential of Perdew (P86) [45]). The table also contains results with the improved approximate xc potential v_{xc}^{SAOP} obtained with the statistical average of (model) orbital potentials (SAOP) [26] (see the next section).

Table 1 shows a gradual improvement in the calculated α along the series $v_{xc}^{LDA} < v_{xc}^{BP} < v_{xc}^{SAOP} < v_{xc}$. The LDA leads, for these relatively "hard" systems, to a polarizability that is systematically too high, which is related to the excess energies of the occupied orbitals, see below. Note, that the GGA-BP gradient correction of the xc potential produces only a relatively small reduction in the LDA/ALDA average absolute error from 8.8% to 5.6%. The improved SAOP potential reduces the error substantially to 2.9%. Still, further significant improvement is achieved with the accurate xc potential; the error from the combination (accurate v_{xc})/ALDA is only 1.0% (See Table 1). Therefore, the crucial improvement of the TDDFRT results for the molecules considered is achieved with just an alteration of the xc potential, while keeping ALDA for the xc kernel.

On the other hand, development of the xc kernel beyond ALDA does not necessarily bring better results. To illustrate this point, the singlet-singlet excitation energies ${\omega_k}^{S}$ calculated in [24] for the He atom with the combination (accurate v_{xc})/ALDA are compared in Table 2 with ${\omega_k}^{S}$ calculated in [46] with the accurate v_{xc} and with the spatially non-local kernel $f_x^{\sigma\sigma(\text{TDOEP})}(\mathbf{r}_1, \mathbf{r}_2)$, which is a part of the exchange-only kernel of the time-dependent optimized effective potential method (OEP) [47]. They are also compared with the TDDFRT zero-order estimate ${\omega_k}^{S0}$, the difference (10) of the orbital energies obtained with the accurate v_{xc}.

Table 2 Helium singlet excitation energies (in Hartrees)

Transition	Exact	KS ω_k^{S0}	TDOEP-SPA	ALDA
$1s\rightarrow 2s$	0.5778	0.7460	0.7687	0.7608
$1s\rightarrow 3s$	0.8425	0.8392	0.8448	0.8435
$1s\rightarrow 4s$	0.8701	0.8688	0.8710	0.8706
$1s\rightarrow 5s$	0.8825	0.8819	0.8830	0.8828
$1s\rightarrow 6s$	0.8892	0.8888	0.8894	0.8893
$1s\rightarrow 2p$	0.7799	0.7772	0.7850	0.7751
$1s\rightarrow 3p$	0.8486	0.8476	0.8500	0.8479
$1s\rightarrow 4p$	0.8727	0.8722	0.8732	0.8724
$1s\rightarrow 5p$	0.8838	0.8836	0.8841	0.8837
$1s\rightarrow 6p$	0.8899	0.8898	0.8901	0.8898
Mean absolute error (mHartrees)	–	2.2	2.2	1.0

ω_k^{S0} refers to the Kohn-Sham orbital energy differences; TDOEP-SPA: exchange only optimized effective potential calculations with the single-pole approximation, see [46]; ALDA: calculations with accurate v_{xc} and the adiabatic local-density approximation for the f_{xc} kernel, see [24]

In this case, from Table 2, one can see that the zero-order estimate ω_k^{S0} with the accurate v_{xc} already yields fair excitation energies, especially for higher excitations. While the addition of the coupling matrix in (9) with $f_{xc}^{\sigma\tau(\mathrm{ALDA})}$ of (44) in general improves the ω_k^{S} obtained, in this case the average absolute error for ω_k^{S} remains unchanged compared with the zero order in calculations where an exchange-only approximation to f_{xc} was used, which was calculated within an optimized effective potential scheme (in other words $f_x^{\sigma\sigma(\mathrm{TDOEP})}$, see column TDOEP-SPA of Table 2). Note that calculations with $f_x^{\sigma\sigma(\mathrm{TDOEP})}$ were performed in [46] in the single-pole approximation (SPA), which effectively means that any contribution of the continuum of unbound states is omitted. In contrast, calculations with $f_{xc}^{\sigma\tau(\mathrm{ALDA})}$ (see column ALDA) were performed in [24] with all KS orbitals resulting from the basis set taken into account, which effectively results in a partial representation of the continuum states. This difference in the basis notwithstanding, the results of Table 2 appear to indicate that the improvement of the form of the xc kernel is not crucial to improving the results of TDDFRT calculations.

Therefore, the main route to better performance of TDDFRT is through improved approximation of the ground-state xc potential v_{xc}, while the simple ALDA for the xc kernel can be used for many chemical applications. Improved approximations for v_{xc} will be considered in the next section.

2.4
Improved xc Potentials

As has been established in the TDDFT literature [13, 17, 22, 38, 48, 49], the main source of errors in the LDA and standard GGA potentials is their considerable upward shift in the bulk atomic and molecular regions compared to the accurate v_{xc} and their improved approximations. Figure 1 illustrates this point by comparing the LDA v_{xc}^{LDA} and GGA-BP v_{xc}^{BP} xc potentials for the N_2 molecule with the SAOP potential v_{xc}^{SAOP} [26] and the potential $v_{xc}^{BP-GRAC}$ which is the BP potential improved by a gradient-regulated asymptotic connection procedure (BP-GRAC) [17, 49] (see also the description of the improved potentials below). At large distances from the N nuclei, all potentials approach the zero long-range asymptotic $v_{xc}(\mathbf{r}_1)\rightarrow 0$, $|\mathbf{r}_1|\rightarrow\infty$. However, in the bulk molecular region the LDA/BP potentials are substantially shifted upward compared to the SAOP/BP-GRAC ones (see Fig. 1).

Due to this shift, the LDA/GGA energies ϵ_i of the occupied orbitals do not satisfy the KS analogue of Koopmans' theorem [50, 51], which interprets the KS orbital energies as the approximate vertical ionization potentials (VIPs)

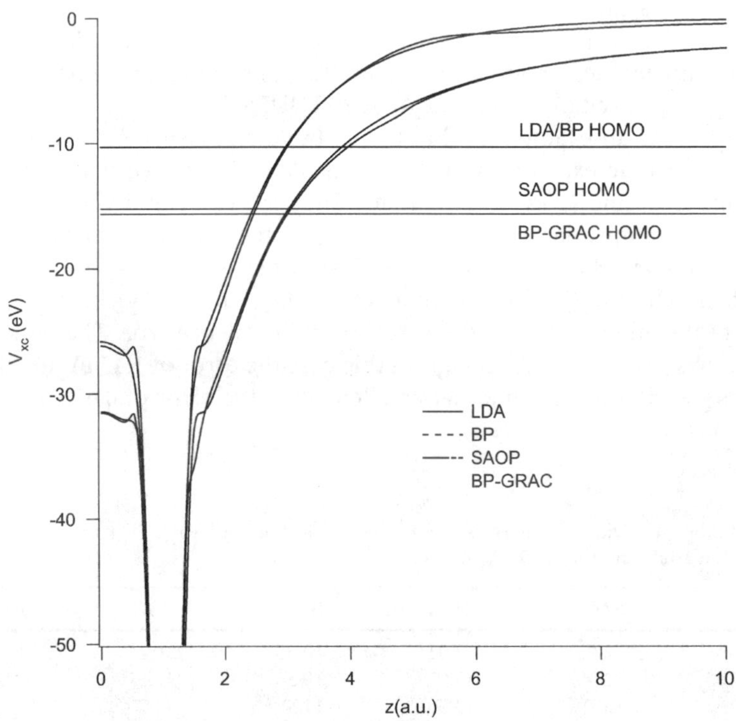

Fig. 1 Plots of the SAOP and BP-GRAC xc potentials for the N_2 molecule. LDA, BP, and BLYP xc potentials are also plotted for comparison

I_i,$-\epsilon_i \approx I_i$. The quality of this approximation appears to be high for valence orbitals, and it becomes the exact identity for the highest occupied molecular orbital (HOMO). Unlike this, the LDA/GGA energies of valence orbitals $-\epsilon_i$ are substantially smaller than the corresponding VIPs. For example, for the HOMO of the N_2 molecule $-\varepsilon_{\mathrm{HOMO}}^{\mathrm{LDA}} = 10.41\,\mathrm{eV}$ and $-\varepsilon_{\mathrm{HOMO}}^{\mathrm{BP}} = 10.39\,\mathrm{eV}$ are smaller by 5.2 eV than the first VIP I_p=15.58 eV.

This deficiency in the LDA/GGA has clear implications for the calculated excitation energies through the TDDFRT zero order $\omega_k^0 = \varepsilon_a - \varepsilon_i$ of (10). For valence-type excitations, the above-mentioned LDA/GGA potential shift does not affect the $\omega_k{}^0$ very strongly. This is because in this case the occupied φ_i and virtual φ_a orbitals are both localized in the bulk region. Due to this, their LDA/GGA energies ϵ_i and ϵ_a will have similar upward shifts, which will cancel each other in the orbital energy difference (10). This explains the rather decent estimates of valence excitations which were obtained within TDDFRT with LDA/GGA.

However, for Rydberg-type excitations the situation will be rather different. In this case the Rydberg-type virtual orbital φ_a has its main amplitude in the outer region, where both LDA/GGA and the accurate potentials are vanishing (though in different ways, see below). Because of this, its energy ϵ_a will have a much smaller shift in LDA/GGA. Due to the presence of the upward shift in the energy ϵ_i of the occupied orbital, this will lead to an underestimated $\omega_k{}^0$ and, as a result, to substantial underestimation of Rydberg-type excitations with LDA/GGA-TDDFRT.

To provide an example of this trend, Table 3 presents the five lowest allowed electronic excitations of the N_2 molecule calculated within TDDFRT with the LDA and GGA-BP potentials. They are compared with the experimental excitation energies. In agreement with the analysis of this section, BP underestimates the energies of the first three excitations by 1.7–2.5 eV, which are all of the Rydberg type. On the other hand, BP yields a good estimate of the other two excitations, which are of valence type. The same trend is also observed for LDA, though in this case the error of 1.45 eV for the last Rydberg excitation is somewhat smaller, while the error of 0.61 eV is larger than those of BP (see Table 3).

Table 3 Lowest excitation energies for the N_2 molecule (in eV) [17] with various Kohn-Sham potentials and the ALDA f_{xc} kernel

Transition	SAOP	BP-GRAC	BP	LDA	Expt
$3\sigma_g \rightarrow 3\sigma_u$	12.93	13.01	10.35	10.46	12.9
$3\sigma_g \rightarrow 2\pi_u$	12.95	13.07	10.36	10.51	12.8
$1\pi_u \rightarrow 4\sigma_g$	13.19	13.35	11.49	11.75	13.2
$2\sigma_u \rightarrow 1\pi_g$	13.58	13.49	13.36	12.90	13.6
$1\pi_u \rightarrow 1\pi_g$	14.08	14.15	14.27	14.27	14.2

Another well-known drawback of the LDA and GGA potentials is their asymptotic behavior. They decay faster than the Coulombic asymptotic behavior $v_{xc}(\mathbf{r}_1)\rightarrow 1/|\mathbf{r}_1|$, $|\mathbf{r}_1|\rightarrow\infty$ required for the accurate xc potential. In the bulk region the LDA and GGA potentials lack the pronounced atomic shell structure of the accurate potential. The improved potentials should possess these features as well as the proper depth in the bulk region, so they should be shifted downward compared to the LDA/GGA potentials.

The first improved xc potential, which found an application in TDDFT, was the potential of van Leeuwen and Baerends (LB94) [38]. The proper downward shift and Coulombic asymptotics are produced in LB94 with the Becke-type [44] gradient correction on top of the LDA potential

$$v_{xc\sigma}^{LB}(\mathbf{r}_1) = v_{xc\sigma}^{LDA}(\mathbf{r}_1) - \frac{\beta x_\sigma^2(\mathbf{r}_1)\rho_\sigma^{1/3}(\mathbf{r}_1)}{1+3\beta x_\sigma(\mathbf{r}_1)\ln\left(x_\sigma(\mathbf{r}_1)+\sqrt{x_\sigma^2(\mathbf{r}_1)+1}\right)}, \tag{45}$$

where x_σ is the dimensionless argument of the density gradient, $x_\sigma(\mathbf{r}_1) = |\nabla\rho_\sigma(\mathbf{r}_1)|/\rho_\sigma^{4/3}(\mathbf{r}_1)$, and β=0.05. The more elaborate SAOP potential $v_{xc\sigma}^{SAOP}$ [26] employs a modified LB potential $v_{xc\sigma}^{LBa}$ as one of its components. The main idea of SAOP is to use different model potentials $v_{xci\sigma}^{mod}$ for the different occupied KS orbitals $\phi_{i\sigma}$ and to combine them into a single xc potential with the statistical average over the orbitals

$$v_{xc\sigma}^{SAOP}(\mathbf{r}_1) = \sum_i^{N_\sigma} v_{xci\sigma}^{mod}(\mathbf{r}_1)\frac{|\phi_{i\sigma}(\mathbf{r}_1)|^2}{\rho_\sigma(r_1)}. \tag{46}$$

In the current SAOP version the orbital potentials $v_{xci\sigma}^{mod}$ are obtained with the interpolation

$$v_{xci\sigma}^{mod}(\mathbf{r}_1) = e^{-2(\varepsilon_{N\sigma}-\varepsilon_{i\sigma})^2} v_{xc\sigma}^{LB\alpha}(\mathbf{r}_1) + \left[1-e^{-2(\varepsilon_{N\sigma}-\varepsilon_{i\sigma})^2}\right] v_{xc\sigma}^{GLLB}(\mathbf{r}_1) \tag{47}$$

between $v_{xc\sigma}^{LBa}$, which has the proper Coulombic asymptotics $-1/r$, and the potential $v_{xc\sigma}^{GLLB}$ of Gritsenko, van Leeuwen, van Lenthe, and Baerends (GLLB) [52], which correctly reproduces the atomic shell structure in the inner regions. With (46) and (47), $v_{xc\sigma}^{SAOP}$ provides a balanced approximation to the KS potential $v_{xc\sigma}$ in all regions.

In an alternative approach, the improved xc potentials have been constructed from the LDA/GGA ones by the explicit inclusion of the potential shift, the latter being estimated from the departure ($\epsilon_{HOMO}{}^{LDA/GGA}+I_p$) of the LDA/GGA HOMO energy $-\epsilon_{HOMO}{}^{LDA/GGA}$ from the first ionization potential I_p. In particular, the improved potential $v_{xc}{}^{CCS}$ of Casida, Casida, and Salahub (CCS) [48] has been constructed by shifting downwards the LDA potential $v_{xc}{}^{LDA}$ in the inner region by $-(\epsilon_{HOMO}{}^{LDA}+I_p)$, using the LB potential $v_{xc}{}^{LB}$ in the outer region, and switching from $[v_{xc}{}^{LDA}-(\epsilon_{HOMO}{}^{LDA}+I_p)]$ to

${v_{xc}}^{LB}$ at the point of intersection of these potentials. In turn, the improved potential ${v_{xc}}^{TH}$ of Tozer and Handy [22] has been constructed with the explicit partitioning of the space of a molecule into inner, intermediate, and outer regions. In the inner region ${v_{xc}}^{TH}$ is just the unmodified GGA potential ${v_{xc}}^{HCTH}$ of Hamprecht, Cohen, Tozer, and Handy (HCTH) [53], while in the outer region it is the asymptotically correct Fermi-Amaldi (FA) potential ${v_{xc}}^{FA}=-v_{Coul}/N$ (N is the total number of electrons) shifted upwards by $({\epsilon_{HOMO}}^{LDA}+I_p)$. In the intermediate region ${v_{xc}}^{TH}$ is obtained with linear interpolation between ${v_{xc}}^{HCTH}$ and $[{v_{xc}}^{FA}+({\epsilon_{HOMO}}^{HCTH}+I_p)]$. Therefore, essentially the same relative downward shift of the inner LDA/GGA part with respect to the asymptotically correct outer part has been used in [22, 48]. In the literature both ${v_{xc}}^{CCS}$ and ${v_{xc}}^{TH}$, as well as the earlier LB94 potential, were called "asymptotically corrected" potentials, which, somewhat misleadingly, stresses the effect of the proper asymptotics. In fact, as the analysis of this section shows, the main feature of the corrections is the relative downward shift of the potential in the bulk region.

Note that the improved potentials ${v_{xc}}^{CCS}$ and ${v_{xc}}^{TH}$ both possess discontinuous derivatives at the borders of the inner and outer regions. Unlike this, in [49] a seamless link between the GGA-BP potential ${v_{xc}}^{BP}$ in the bulk region and the shifted LB potential at the asymptotics has been achieved with the gradient regulated asymptotic connection (GRAC)

$$v_{xc}^{BP-GRAC}(\mathbf{r}_1) = [1-f(x)]v_{xc}^{BP}(\mathbf{r}_1) + f(x)\left[v_{xc}^{LB}(\mathbf{r}_1) + \left(\varepsilon_{HOMO}^{BP} + I_p\right)\right]. \tag{48}$$

In the improved GRAC-BP potential of (48) $f(x)$ is the gradient-dependent interpolation function

$$f[x(\mathbf{r}_1)] = \frac{1}{1+e^{-\alpha[x(\mathbf{r}_1)-\beta]}}, \tag{49}$$

where

$$\mathbf{x}(\mathbf{r}_1) = \nabla\rho(\mathrm{r}_1)/\rho^{4/3}(\mathrm{r}_1)$$

with α=0.5, β=40.

To illustrate the effect of the improvement of potentials, Table 3 also presents the five lowest allowed electronic excitations of the N_2 molecule calculated within TDDFRT with the SAOP and GRAC-BP potentials [17]. One can see from Table 3 that both potentials provide excellent TDDFRT estimates of the excitations considered. Both SAOP and GRAC-BP virtually remove the large LDA/BP errors for the Rydberg-type excitations, while keeping a rather high accuracy for the valence excitations. Especially pleasing is the performance of the SAOP, which provides the proper form of the potential naturally, without explicit shift $-({\epsilon_{HOMO}}^{LDA}+I_p)$ of its inner part. Therefore, while ${v_{xc}}^{CCS}$, ${v_{xc}}^{TH}$, and ${v_{xc}}^{BP\text{-}GRAC}$ all require the experimental or precalculated ionization potentials I_p, SAOP is free from this requirement.

TDDFRT presented in this section is also applied within the time-dependent hybrid approach. It parallels the corresponding approach in DFT and it combines TDDFRT with the time-dependent Hartree-Fock (TDHF) theory [10, 54]. Instead of a pure DFT xc potential $v_{xc\sigma}$, the hybrid approach employs for the orbital $\phi_{i\sigma}$ in (7) an admixture of an approximate potential $v_{xc\sigma}$ with the exchange Hartree-Fock potential $v_{xi\sigma}$ for this orbital

$$v_{xi\sigma} = -\frac{1}{\phi_{i\sigma}^*(\mathbf{r}_1)} \sum_{j=1} \phi_{j\sigma}^*(\mathbf{r}_1) \int \frac{\phi_{i\sigma}^*(\mathbf{r}_2)\phi_{j\sigma}(\mathbf{r}_2)}{r_{12}} d\mathbf{r}_2 \tag{50}$$

With this, the excitation energies ω_k are calculated in the hybrid scheme via (7) and (9) where the matrix element $K^{\text{xc}}_{ij\sigma,kl\tau}$ of the pure kernel is replaced with a mixture of an approximate $K^{\text{xc}}_{ij\sigma,kl\tau}$ with the corresponding TDHF element $K^{(x)\text{HF}}_{ij\sigma,kl\tau}$

$$K^{(x)\text{HF}}_{ij\sigma,kl\tau} = -S_{\sigma\tau} \int \phi_{i\sigma}^*(\mathbf{r}_1)\phi_{k\sigma}(\mathbf{r}_1)\frac{1}{r_{12}}\phi_{j\tau}(\mathbf{r}_2)\phi_{l\tau}^*(\mathbf{r}_2)d\mathbf{r}_1 d\mathbf{r}_2 \tag{51}$$

TDDFRT/TDHF mixture coefficients, as well as the composition of the approximate $v_{xc\sigma}$, depend on the hybrid scheme used. Applications of the TDDFRT/TDHF approach with the hybrid functionals, such as the hybrid Becke's three-parameter functional [55] and the gradient correction of Perdew, along with his 1981 local correlation functional [45], B3P86, the Lee–Yang–Parr nonlocal correlation functional [56], B3LYP, or the Perdew–Burke–Erzenrhof (PBE) [57, 58] hybrid functional, PBE0 [59], will be discussed below.

The behavior of the various potentials in the calculation of excitation energies of transition metal complexes will be considered in the next sections.

3 Organometallic Complexes

3.1 Metallocarbonyls

Transition metal carbonyls represent a very important class of complexes because of their high chemical and photochemical reactivity. They are among the most reactive transition metal complexes and their electronic structure has long been a matter of considerable interest. Most of the theoretical interest has for a long time centered on an understanding of the bonding mechanism. Special attention has been given to the intriguing question of the relative importance of σ-donation and π-back donation in the synergic M–CO bond. A comparable intensity of effort has been devoted to

Table 4 TDDFT/BP excitation energies (eV) and oscillator strengths (in parentheses) of the lowest allowed excited states of $Ni(CO)_4$, compared to CASPT2 and SAC-CI values and to experimental data

State	TDDFT [68]	CASPT2 [70]	SAC-CI [72]	Expt
1^1T_2	4.70 (0.006)	4.34 (0.29)	4.79 (0.0023)	4.5[a]; 4.54[b]
2^1T_2	4.82 (0.099)	5.22 (0.38)	5.51 (0.0067)	5.4[a]; 5.17[b]; 5.24[c]
3^1T_2	5.37 (0.123)	5.57 (0.29)	5.72 (0.0043)	5.52[c]
4^1T_2	5.84 (0.086)	6.28 (0.47)	5.76 (0.0089)	6.0[a]; 6.02[c]
5^1T_2	6.74 (1.002)	6.97 (0.83)		

[a] Gas-phase spectrum from [65]; [b] matrix spectrum from [61]; [c] solution spectrum from [76]

the study of metal–metal bonding with the aim of rationalizing the stereochemical variety and isomeric richness exhibited by metal–carbonyl dimers, trimers, and clusters. Somewhat remarkably however, relatively little theoretical attention has been given to the electronic spectroscopy of these complexes. Their experimental UV spectra have been known since the early 1970s [60–65], and until the mid 1990s, when the first calculations based on correlated ab initio methods began to appear, only tentative assignments were given, mostly based on the earliest interpretation of the spectra by means of qualitative molecular orbital schemes. The successes of DFT methods in describing the ground state properties of transition metal complexes, including metallocarbonyls, in combination with the reliability of the TDDFT approach in obtaining accurate predictions of excitation energies and oscillator strengths for higher fullerenes [66] and large organic molecules [10, 23, 67], made it attractive to apply TDDFT to the calculations of the electronic spectra of metallocarbonyls. The electronic spectra of the binary carbonyl compounds, $Ni(CO)_4$ [68], $M(CO)_6$ (M=Cr, Mo, W) [69], and of the binuclear metal–metal bonded carbonyl, $Mn_2(CO)_{10}$ [68], have been investigated so far by TDDFT methods. The results are reported in Tables 4, 5, and 6 and compared to the experimental values as well as to high quality ab initio calculations, such as CASPT2 [70, 71] and symmetry adapted cluster configuration interaction (SAC–CI) [72]. In addition, the TDDFT results are compared to excitation energy calculations in which an older DFT approach was used, the ΔSCF-DFT approach [68, 73–75].

Experimentally, the spectrum of $Ni(CO)_4$ has been measured in solution [76], in matrix [61] and, more recently, in gas-phase [65]. The solution spectrum exhibits a main peak at 6.0 eV, with two shoulders at 5.5 and 5.2 eV. Only the 5.2 eV band could be detected in the matrix isolation spectrum, with a new shoulder appearing at 4.5 eV. Finally, in the gas-phase spectrum a main peak at 6.0 eV and two shoulders at 5.4 and 4.6 eV could be discerned. Being a d^{10} system, $d \rightarrow d$ ligand field (LF) transitions are absent in the spectrum. Since $Ni(CO)_4$ has T_d symmetry, only excitations from the $^1A_1$

Table 5 TDDFT excitation energies (eV) and oscillator strengths (in parentheses) of the lowest allowed excited states of $M(CO)_6$ (M=Cr, Mo, W) are compared to the experimental data, and, in the case of $Cr(CO)_6$, to ΔSCF-DFT and CASPT2 results

State	ΔSCF-DFT [74]	TDDFT [69]		CASPT2 [70][a]	Expt [60]
		BP	LB94		
$Cr(CO)_6$					
1^1T_{1u}	5.60	4.19 (0.03)	3.91	4.54–4.11 (1.33–0.20)	4.43 (0.25)
2^1T_{1u}	6.50	5.76 (1.52)	5.37	5.07–5.20 (1.63–2.58)	5.41(2.30)
$Mo(CO)_6$[b]					
1^1T_{1u}		3.88 (0.04)	3.60		4.33 (0.15)
2^1T_{1u}		5.73 (2.16)	5.40		5.45 (2.20)
$W(CO)_6$[b]					
1^1T_{1u}		3.80 (0.04)	3.55		4.30 (0.18)
2^1T_{1u}		5.84 (2.41)	5.52		5.53 (3.30)

[a] The energy and oscillator strength ranges refer to different choices of active spaces, see [70] for details; [b] relativistic effects are included at scalar relativistic (SR) ZORA (zero-order regular approximation) level [192–196], see [69] for details

ground state to excited $^1T_2$ states are allowed. There are five such MLCT excitations from the occupied $3d$ states (the 9 t_2 and 2e orbitals) to the ligand manifold of $2\pi^*$ orbitals. The experimental spectra however exhibit at most three spectral bands in all cases [61, 65, 76]. The interpretation of the $Ni(CO)_4$ spectrum has therefore initially been restricted to the general characterization of the spectral bands as MLCT transitions. The first theoretical study of the spectrum used the intermediate neglect of differential overlap in the spectroscopic parameterization model with a subsequent configuration interaction (INDO/S CI) [65]. The agreement with the experiment was poor and the assignment given was in disagreement with those based on higher level approaches (SAC-CI [72], CASPT2 [70], and TDDFT/BP [68]).

Table 6 TDDFT/BP excitation energies (eV) and oscillator strengths (in parentheses) of the lowest allowed excited states of $Mn_2(CO)_{10}$, compared to experimental data and to ΔSCF-DFT and CASPT2 results

State	ΔSCF-DFT [68][a], [73][b]		TDDFT/BP [68][a]		CASPT2 [71][c]		Expt[d]
1^1E_1	3.07[a]; 3.62[b]	$d_\pi \rightarrow \sigma^*$	3.22(0.007)[a]; 3.44(0.006)[c]; 3.35(0.006)[e]	$d_\pi \rightarrow \sigma^*$	3.29(0.03)	$d_\pi \rightarrow \sigma^*$	3.31
1^1B_2	2.95[a]; 3.42[b]	$\sigma \rightarrow \sigma^*$	3.76(0.384)[a]; 4.01(0.252)[c]; 3.84(0.350)[e]	$\sigma \rightarrow \sigma^*$	3.43(0.92)	$\sigma \rightarrow \sigma^*$	3.69

[a] Becke-Perdew optimized geometry (see [81], Table 6, row 9); [b] Becke-Perdew optimized geometry; [c] X-ray structure from [82]; [d] solution spectrum of $Mn_2(CO)_{10}$ from [63]; [e] electron diffraction structure from [83]

The TDDFT excitation energy and oscillator strength computed for the five $^1T_2$ MLCT excited states of $Ni(CO)_4$ are gathered in Table 4 together with the experimental data and the best CASPT2 and SAC-CI results. The TDDFT, CASPT2, and SAC-CI approaches all give rise to four $^1T_2$ excited states in the energy window 4.0–6.5 eV, in agreement with experiment. They come however to a quite different assignment of the spectrum. CASPT2 assigns the main peak at 6.0 eV in the solution and gas-phase spectra to the 9 $t_2 \rightarrow 2\ t_1$ MLCT transition, which is the 4^1T_2 state in that calculation, and the two shoulders observed at 5.2 and 5.5 eV in the solution spectrum to the 9 $t_2 \rightarrow 3e$ and $2e \rightarrow 10\ t_2$ MLCT transitions respectively, which are the 2^1T_2 and 3^1T_2 excited states in the CASPT2 calculations. Finally, the lowest and least intense band observed at 4.5–4.6 eV in the matrix isolation and gas-phase spectrum is attributed to the lowest calculated excited state corresponding to the 9 $t_2 \rightarrow 10\ t_2$ MLCT transition. TDDFT agrees with CASPT2 on the assignment of the lowest four states as $^1T_2$ MLCT states, although the composition of the third and fourth state is reversed in the two approaches. In the SAC-CI study the three lowest $^1T_2$ MLCT states were computed at 4.79, 5.51, and 5.76 eV (labeled as 1^1T_2, 2^1T_2, and 4^1T_2). In the same energy regime, a fourth $^1T_2$ state, the 3^1T_2, corresponding to a Ni $3d \rightarrow 4s$ transition, was calculated at 5.72 eV. Based on these SAC-CI results, a different assignment of the spectrum was proposed, at least for the bands at 5.4 and 6.0 eV. The former was assigned to the 2^1T_2 MLCT state, the latter to the 3^1T_2 $3d \rightarrow 4\ s$ and 4^1T_2 MLCT nearly degenerate states. CASPT2, SAC-CI and TDDFT agree on the assignment of the lowest MLCT state to the weak feature observed at 4.5–4.6 eV in the matrix isolation and gas-phase spectrum.

As regards the calculated oscillator strengths, the CASPT2 (based on CASSCF dipole moment integrals) and SAC-CI results differ by almost two orders of magnitude, while TDDFT results are somewhere in-between. It is apparent however from the values in Table 4 that none of the three approaches puts the highest intensity into the main peak of the spectrum at 6.0 eV. The largest oscillator strength is actually predicted by both CASPT2 and TDDFT for the 5^1T_2 state at the considerably higher energy of 6.74 (TDDFT) or 6.97 (CASPT2). Although the three approaches in Table 4 give reasonable agreement with experiment and each other, it is clear that further theoretical work is needed in order to give definitive, unambiguous assignments and oscillator strengths, in particular for the features above 4.5 eV.

The electronic spectra of metal hexacarbonyls, $M(CO)_6$ (M=Cr, Mo, W), are very similar to each other [60, 77]. They are dominated by two very intense absorption bands, which can be identified as the only two orbitally- and spin-allowed $^1A_{1g} \rightarrow {}^1T_{1u}$ MLCT excitations, and by low-lying shoulders originally assigned by Gray and Beach [77] as vibrational components of the ligand field (LF) excited state $^1T_{1g}$ belonging to the $t_{2g}{}^5e_g{}^1$ configuration. Although the authors themselves were concerned about this assignment because it suggests the LF splitting is the same for chromium, molybdenum and tungsten carbonyls, it appeared to be confirmed by the original extended-Hückel [77] as well as, in the case of $Cr(CO)_6$, semiempirical INDO/S CI [65] and ab initio restricted Hartree-Fock (RHF) [78] calculations. There was also little reason for revision of this assignment, since irradiation in the low-energy shoulder leads to fast ejection of CO, in perfect agreement with the "standard" model of photodissociation being induced by LF excitation [79, 80]. Relativistic TDDFT calculations by Rosa et al [69] of the excited states of the $M(CO)_6$ (M=Cr, Mo, W) series, in agreement with previous ΔSCF-DFT [74] and CASPT2 [70] calculations on $Cr(CO)_6$, have shown that in all members of the series the low-energy shoulders in the spectra do not correspond to LF excitations, as had been accepted for a long time. Instead they correspond to orbitally forbidden MLCT states. The LF excitations are calculated at much higher energy than suggested by the original assignment and at different energy along the $M(CO)_6$ series, being much higher in the heavier carbonyls than in $Cr(CO)_6$. These results have led to a definitive reassessment of the role of the LF states in the photochemical dissociation of the metal-CO bond in the $M(CO)_6$ series, suggesting that the experimentally observed photodissociation of the M–CO bond upon irradiation into the lowest energy band occurs in the heavier carbonyls, as it does in $Cr(CO)_6$, from CT and not from LF states.

As inferred from Table 5, the two intense bands of the spectra of $M(CO)_6$, corresponding to the two allowed CT transitions, are quite well accounted for by TDDFT excitation energies. In the case of $Cr(CO)_6$ there is a substantial and general agreement between CASPT2 and TDDFT results. The improvement of TDDFT over ΔSCF-DFT results (also in cases not reported here) is considerable for all computed excited states, but it is especially noteworthy for the two dipole allowed $^1T_{1u}$ states reported in Table 5, which at the ΔSCF-DFT level are computed to be more than 1 eV higher than the experimental band maxima. The particularly poor ΔSCF-DFT results for the 1^1T_{1u} and 2^1T_{1u} are attributed to the inability of this approach to account for the configurational mixing occurring in these states which have a multiple transition character, consisting of the same two transitions, the $2t_{2g}\rightarrow 9t_{1u}$ and $2t_{2g}\rightarrow 2t_{2u}$, with approximately reversed weights. Comparison between the BP and LB94 results of [69] reveals that the choice of the xc functional has a large effect on both singlet and triplet states, the LB94 results being usually, but not invariably, closer to CASPT2 than the BP ones. In the case of the two $^1T_{1u}$ states (see Table 5) comparison with experiment and the CASPT2 results gives a mixed picture. The LB94 excitation energy is too low for the 1^1T_{1u}, and just right for the 2^1T_{1u}, while the reverse holds for the CASPT2 values. In the heavier carbonyls, the effects of the choice of the xc functional on the excitation energies are very similar to those observed in $Cr(CO)_6$, the LB94 excitation energies being in general lower than BP ones by 0.2–0.5 eV. As in $Cr(CO)_6$, the LB94 excitation energy is too low for the 1^1T_{1u} and just right for the 2^1T_{1u}. According to Rosa et al [69], the sensitivity to the choice of the potential of the excited states computed for the $M(CO)_6$ series, which are diffuse (many excited states involve transitions to pure CO $2\pi^*$ orbitals), but in most cases not at all Rydberg-like, indicates that not only the asymptotic behavior of the potential is an important factor, but that the behavior in the region with important electron density has to be taken into account as well. This is underlined by the effects for the LF excited states (not shown here) which show an energy shift of ~0.2 eV on going from BP to LB94, despite the nominally *d-d* character of these excitations. As for the Rydberg excited states, they are calculated at the LB94 level to be 1 eV *higher* than at the BP level.

$Mn_2(CO)_{10}$ has attracted considerable interest, mostly because of its rich photochemistry that is generally taken as a prototype for photoreactions of organometallic compounds containing metal–metal bonds. Two competitive primary photoreactions occur either in the condensed-phase or in the gas-phase photolysis: Mn–Mn bond cleavage and Mn–CO bond dissociation. The quantum yield of these two reactive channels is controlled by the excitation wavelength. Since the early work by Gray and coworkers in the 1970s [62], numerous experimental and theoretical studies have been dedicated to the spectral and photochemical properties of this prototype dinuclear metallocarbonyl. The main focus was on the nature of the low-lying excited states and their role in the photodissociation processes. The lowest part of the absorption spectrum of $Mn_2(CO)_{10}$ is characterized by a broad and relatively intense band centered at 3.69 eV (Band II) and a poorly resolved shoulder to the red side of this band, at 3.31 eV (Band I). On the basis of a variety of experimental and semiempirical theoretical data, Gray et al [62, 63] assigned the weak Band I to the $d_\pi \rightarrow \sigma*$ transition ($1^1A_1 \rightarrow 1^1E_1$ in the pertinent D_{4d} symmetry of the molecule) and the strong Band II to the $1^1A_1 \rightarrow 1^1B_2$, $\sigma \rightarrow \sigma^*$ transition (σ and σ^* denote the Mn–Mn bonding and antibonding orbitals respectively). This early assignment by Gray et al has been confirmed recently by TDDFT [68] and CASPT2 [71] calculations whose results are, as inferred from Table 6 in excellent agreement with the experiment. By contrast, ΔSCF-DFT calculations [68, 73], which show a strong dependence on the geometry used, predict the bands I and II in the reverse order. The dependence of the excitation energies on the geometry of the TM complexes will be a recurrent theme in this review. In Table 6 TDDFT results are reported for three different geometries: the BP optimized geometry obtained by Rosa et al [81] using a large basis set (QZDF for C and O and TZ3P for Mn), the experimental geometries obtained by X-ray diffraction at room temperature, and by electron diffraction at 74 K [83]. The differences underline the sensitivity of the calculated excitation energies (and oscillator strengths) to the molecular structure.

3.2 Metal Sandwich Compounds

TDDFT methods have been used to interpret the electronic spectroscopy of bis(cyclopentadienyl) and mono(cyclopentadienyl) transition metal complexes. A detailed analysis of the dependence of the excitation energies on the exchange-correlation functional and on the ground state geometry has been reported by Boulet et al [84] for two bis(cyclopentadyenyl) complexes, $FeCp_2$ and $RuCp_2$, and for the mono(cyclopentadienyl) CpNiNO system. The absorption spectra of ferrocene and ruthenocene are characterized by two weak metal-centered (MC) bands, Band II and Band III, around 2.7 and 3.8 eV in $FeCp_2$, and 3.84 and 4.54 eV in $RuCp_2$. Band II splits at low temperature (77 K) into two components with maxima at 2.70 and 2.98 eV in $FeCp_2$, and 3.66 and 4.03 eV in $RuCp_2$. The MLCT bands appear at much higher energy and are much more intense. According to the TDDFT results gathered in Table 7, the two maxima of Band II in $FeCp_2$ are accounted for by the two *d–d* transitions $^1E_{1g}$ ($5a_{1g}\rightarrow 4e_{1g}$) and $^1E_{2g}$ ($3e_{2g}\rightarrow 4e_{1g}$), whereas Band III originates from a single *d–d* transition, $^1E_{1g}$ ($3e_{2g}\rightarrow 4e_{1g}$). The observed excitation energies are satisfactorily reproduced using TDDFT with the asymptotically correct LB94 and LRC [85] functionals, whereas they are largely underestimated by the B3LYP functional. Table 7 also illustrates the sensitivity of the calculated excitation energies to the geometry. LB94 results are given for three different geometries, obtained by geometry optimization with different functionals. The excitation energies often differ by as much as 0.4 eV. The LDA (VWN) geometries are closest to experiment, particularly for $RuCp_2$, and indeed on average LB94/LDA gives the closest agreement with experiment. For an extensive analysis of the influence of basis set and exchange-correlation functional on the accuracy of the geometries of cyclopentadienyl complexes, the reader is referred to the recent paper by Swart and Snijders [86]. ΔSCF-DFT [87] reproduces satisfactorily the observed bands of both $FeCp_2$ and $RuCp_2$ and gives the correct assignment of the experimental spectra. As for the very weak spin forbidden *d–d* absorption (Band I) observed in the spectra of these metallocenes (at 2.34 eV in $FeCp_2$ and 3.22 eV in $RuCp_2$), it is nicely accounted for by LB94/LDA results, which assign this band to the $^3E_{2g}$ ($3e_{2g}\rightarrow 4e_{1g}$) in $FeCp_2$ and to the $^3E_1''$ ($5a_1'\rightarrow 4e_1''$) in $RuCp_2$.

According to TDDFT results by Boulet et al [84], for the CpNiNO complex, unlike for metallocenes, the BP functional improves the description of both the geometrical parameters and the excited states.

A careful analysis, based on CASSCF/CASPT2 and TDDFT calculations, of the electronic spectrum of η_5-$CpMn(CO)_3$, a mono(cyclopentadienyl) complex which has been widely investigated for its photochemical reactivity, has been reported by Full et al [88]. The capability of TDDFT and highly correlated ab initio methods to describe MC and MLCT states of this organome-

Table 7 Observed and calculated excitation energies (eV) for the MC transitions of ferrocene and ruthenocene complexes

Molecule	State	Character	ΔSCF-DFT [87]	TDDFT [84]					
				LB94/LDA[a]	LB94/BP[b]	LB94/PW91[c]	LRC/BP[b]	B3LYP[d]	Expt [197]
$FeCp_2$	$^3E_{2g}$	$3e_{2g}\rightarrow 4e_{1g}$	–	2.304	1.890	1.92	1.87		2.34 (Band I)
	$^1E_{1g}$	$5a_{1g}\rightarrow 4e_{1g}$	–	3.092	2.745	2.63	2.56	2.27	2.70 (Band II)
	$^1E_{2g}$	$3e_{2g}\rightarrow 4e_{1g}$	2.87 ($^1E_2''$)	2.975	2.561	2.41	2.35	1.74	2.98 (Band II)
	$^1E_{1g}$	$3e_{2g}\rightarrow 4e_{1g}$	3.48 ($^1E_1''$)	3.621	3.245	3.07	3.16	3.22	3.82 (Band III)
	r_{Fe-Cp} (Å)[e]			1.637	1.693	1.689	1.693	–	1.660
$RuCp_2$	$^3E_1''$	$5a_1'\rightarrow 4e_1''$	3.33	3.34	2.99	3.01	2.99	–	3.22 (Band I)
	$^1E_1''$	$5a_1'\rightarrow 4e_1''$	3.51	3.69	3.35	3.37	3.34	–	3.66 (Band II)
	$^1E_2''$	$3e_2''\rightarrow 4e_1''$	3.97	3.80	3.39	3.42	3.37	–	4.03 (Band II)
	$^1E_1''$	$3e_2''\rightarrow 4e_1''$	–	4.21	3.82	3.85	3.78	–	4.54 (Band III)
	r_{Ru-Cp} (Å)[e]			1.810	1.859	1.856	1.859	–	1.816

[a] Energies calculated at DFT/LDA/TZP optimized geometries (D_{5d} and D_{5h} symmetry for $FeCp_2$ and $RuCp_2$ respectively); [b] energies calculated at DFT/BP/TZP optimized geometries; [c] energies calculated at DFT/PW91/TZP optimized geometries; [d] energies calculated at DFT/B3LYP/LAND2DZ optimized geometries; [e] r_{M-Cp} is the vertical M–ring distance

Table 8 Excitation energies (eV) and oscillator strengths (in parentheses) of the low-lying $1^1A'$ excited states of η_5-CpMn(CO)$_3$ calculated at the CASSCF/MS-CASPT2 and TDDFT theoretical levels [88]

State	CASPT2[a]	TDDFT[b]		Expt [198]
		B3LYP	BP	
MC states				
b^1A'	3.19 (0.0069)	3.31 (0.0097)	3.44 (0.0114)	(3.31 onset)
c^1A'	3.28 (0.0157)	3.38 (0.0025)	3.58 (0.0035)	3.81
d^1A'	3.76 (0.0015)	3.84 (0.0000)	3.81 (0.0001)	
MLCT states				
e^1A'	4.47 (0.0000)	4.12 (0.0000)	4.25 (0.0005)	(4.50)
f^1A'	5.04 (0.0040)	4.56 (0.0132)	4.64 (0.0014)	
g^1A'	5.11 (0.0052)	4.94 (0.0207)	4.67 (0.0011)	4.96

[a] Energies calculated at the CASSCF optimized geometry (C_s symmetry); [b] energies calculated at the TDDFT/B3LYP optimized geometry (C_s symmetry)

tallic system has been thoroughly tested in this study. In Table 8 the TDDFT results obtained with two different exchange correlation functionals, the hybrid B3LYP and the standard GGA-BP potentials, are compared with CASSCF/CASPT2 results and the experimental band maxima.

The UV spectrum of η_5-CpMn(CO)$_3$ is characterized by a shoulder starting at 375 nm (3.31 eV) with a maximum at 325 nm (3.81 eV), and a second band with a rather high intensity at 275 nm (4.50) eV with a peak around 250 nm (4.96 eV). CASPT2 and TDDFT agree on the assignment of the weak shoulder to the three lowest $^1A'$ excited states corresponding to $3d_{Mn} \rightarrow 3d_{Mn}$ MC transitions, and of the second band to the f^1A' and g^1A' states corresponding to $3d_{Mn} \rightarrow \pi^*_{CO}$ MLCT transitions. The calculated BP and B3LYP excitation energies compare rather well to CASPT2 results as far as the MC states are concerned, although both approaches underestimate the experimental band maximum. As for the MLCT states, the TDDFT energies are lower than the CASPT2 values. Comparison with the experiment reveals however that TDDFT/B3LYP reproduces the experimental band maximum at 4.96 eV (see Table 8) as well as CASPT2. The BP calculations have comparable first and second MLCT excitation energies, but the third is a bit low (4.67 eV) compared to the experimental peak at 4.96 eV. We note, however, that the excitations have been calculated at the B3LYP optimized geometry, which is not particularly reliable for TM complexes. A more careful investigation of geometry effects would be needed in order to definitively establish the relative accuracy of B3LYP versus BP (or other DFT methods) for this molecule.

4
Werner Complexes

Several types of Werner complexes have been investigated over the last few years by TDDFT methods. They include metal oxide, metal halide, metal oxyhalide compounds, and transition metal complexes with bidentate ligands such as ethylenediamine and acetylacetonato.

Among metal oxide compounds, the permanganate ion, MnO_4^-, is a very attractive system for theoretical study. The assignment of the electronic spectrum of MnO_4^- has undergone a lot of changes throughout the years (for an overview see [89]) and is still a challenge for quantum chemists. The highest lying occupied orbitals consist of non-bonding oxygen orbitals that split into a_1, t_1, and t_2 representations. The low-lying virtual orbitals are primarily antibonding oxygen-metal *d* orbitals and comprise the t_2 and e representation in addition to a 4s a_1 orbital. The low-lying excitations in permanganate are all charge transfer excitations from non-bonding oxygen to metal-ligand antibonding orbitals. However, the bonding metal-oxygen orbitals that lie below the nonbonding oxygen orbitals play an important role. There are substantial correlation effects in the ground state that involve excitations from these bonding O-2*p*/Mn-3*d* orbitals into the virtual antibonding manifold [90]. Moreover, these correlation effects change appreciably in the excited states. All of these orbitals would be required in multireference-based methods like CASPT2 and MRCI, which makes MnO_4^-, as well as the other members of the d^0 series of metal oxides that will be discussed in this section, CrO_4^{2-} and RuO_4, far beyond current computational limits.

The ground and excited states of MnO_4^- have been investigated by a variety of theoretical approaches, SCF-Xα [91], SCF [92], CI(SD) [93], ΔSCF-DFT [94-96], cluster expansion methods, such as SAC-CI [89] and EOM-CCSD [97], including the recently implemented TDDFT methods [68, 84]. The experimental spectrum of MnO_4^- was measured by Holt and Ballhausen [98]. The measured energy maxima are reported in Table 9 together with TDDFT [68, 84], ΔSCF-DFT [96], SAC-CI [89], and EOM-CCSD [97] theoretical results for the spin- and dipole-allowed excitation energies, which belong to the T_2 irreducible representation of this tetrahedral molecule. The data in Table 9 clearly show that TDDFT methods systematically overestimate the transition energies to a variable extent, depending on the functional and on the ground state geometry used. When comparing the BP and LB94 excitation energies obtained by van Gisbergen et al [68] for the experimental geometry, it is apparent that the asymptotically correct functional (LB94) lowers the energies by 0.2–0.4 eV and leads to an improvement of the results, although the first and third excited states still deviate considerably from experiment (0.3–0.5 eV). We stress again the sensitivity of the results to the chosen geometry, as is evident from the LB94 data obtained at different geometries. At the LDA geometry (R(Mn–O)=1.607 Å, experimental

Table 9 Excitation energies (eV) of MnO_4^- computed by TDDFT methods compared to ΔSCF-DFT, SAC-CI, EOM-CCSD, and experimental values

State	ΔSCF [96]	TDDFT BP [68][a]	LB94 [68][a]	LB94/LDA [84][b]	LB94/BP [84][c]	LB94/PW91 [84, 200][b]	LRC/BP [84][c]	B3LYP [84][e]	SAC-CI [89][f]	EOM-CCSD [97][g]	Expt (refs in [])
a^1T_2	2.70	2.82	2.63	2.76	2.63	2.83	2.63	2.81	2.57	2.24	2.27 [98]
b^1T_2	4.01	3.89	3.60	3.77	3.60	3.90	3.63	3.88	3.58	3.67	3.47 [98]
c^1T_2	4.26	4.74	4.52	4.72	4.52	4.76	4.52	4.41	3.72	3.60	3.99 [98]
d^1T_2	5.69	5.84	5.46	5.70	5.47	5.87	5.50	–	5.82	5.80	5.45 [98]
r_{Mn-O} (Å)		1.629	1.629	1.607	1.628	1.626	1.628	1.613	1.629	1.629	1.629 [199]

[a] Energies calculated at the experimental T_d geometry [199]; [b] energies calculated at the DFT/LDA/TZP optimized geometry; [c] energies calculated at the DFT/BP/TZP optimized geometry; [d] energies calculated at the DFT/PW91/TZP optimized geometry; [e] energies calculated at the DFT/B3LYP/LAND2DZ optimized geometry; [f] energies calculated at the experimental T_d geometry [199]; [g] energies calculated at the experimental T_d geometry [199]

1.629 Å), all of the excitation energies shift up by ~0.2 eV, which is consistent with an expected up-shift of the antibonding virtual levels upon shortening of the bond length. This effect is significant. The results by Boulet et al [84] show that at the BP optimized geometry, which is actually very close to the experimental one (1.628 versus 1.629 Å), the two asymptotically corrected functionals, LB94 and LRC, both give results very close to each other, and as expected very close to the LB94 results of van Gisbergen et al [68] at the experimental geometry. However, the LB94 results of Boulet et al [84] at the PW91 optimized geometry, which is still very close to experiment (1.626 Å), deviate remarkably from the other LB94 results at 1.629 and 1.628 Å. It seems unlikely that such a small geometry change can have such large effects (0.2–0.4 eV). The hybrid functional, B3LYP, does not perform particularly well, comparable to BP and definitely worse than LB94.

The cluster expansion based methods, such as SAC-CI and EOM-CCSD approaches, appear to be rather reliable, although still with an occasional deviation from experiment by 0.3–0.4 eV. The ΔSCF-DFT approach is based on orbital replacement energies and is not able to adequately take into account the strong configurational mixings observed at TDDFT level, even though the self-consistent procedure for the excited states should partially account for the mixing between several orbital replacements. As a consequence, not only the energies but also the assignment proposed by the density functional based methods differ significantly. All of the methods gathered in Table 9 agree on the assignment of the first band to an excited state (a^1T_2) with a predominant $1t_1 \rightarrow 2e$ character. TDDFT and SAC-CI agree in describing the fourth band as a mixing of the $6t_2 \rightarrow 7t_2$ and $6a_1 \rightarrow 7t_2$ transitions. However, the assignment of the second and third band is still controversial. The SAC-CI, ΔSCF-DFT, and EOM-CCSD assign the second band to the $1t_1 \rightarrow 7t_2$ transition and the third to the $6t_2 \rightarrow 2e$. On the other hand, TDDFT, in agreement with CI(SD) [93], (not reported in Table 9) come to a basically reversed assignment of these bands, although at TDDFT level the configurations responsible for these bands are quite strongly mixed in the pertinent b^1T_2 and c^1T_2 excited states. It is clear from the theoretical results obtained so far that MnO_4^- remains a quite difficult system.

According to the excitation energies gathered in Table 10, the performance of TDDFT improves sensibly when going from MnO_4^- to CrO_4^{2-} and RuO_4. The absorption spectrum of an aqueous solution of potassium chromate shows two main bands at 3.36 and 4.61 eV and a very weak shoulder around 2.75 eV. TDDFT calculations by Boulet et al have assigned the first band to the a^1T_2 ($1t_1 \rightarrow 2e$) LMCT excited state, and the second band to the b^1T_2 ($4t_2 \rightarrow 2e$) and c^1T_2 ($1t_1 \rightarrow 5t_2$) LMCT states. The results obtained with different functionals and geometries are slightly scattered, but the statistical significance of the prediction is comparable to the experimental accuracy of the room temperature solution spectrum. From a theoretical point of view, the results obtained for the RuO_4 complex are basically similar to those ob-

Table 10 Excitation energies (eV) of CrO_4^{2-} and RuO_4 computed by TDDFT methods compared to ΔSCF-DFT and experimental values

State	ΔSCF [96]	TDDFT [84]					Expt (refs in [])
		LB94/LDA[a]	LB94/BP[b]	LB94/PW91[c]	LRC/BP[b]	B3LYP[d]	
CrO_4^{2-}							
a^1T_1		2.99	2.84	2.85	2.85	3.15	2.75 [200]
a^1T_2	3.62	3.36	3.20	3.21	3.21	3.60	3.36 [200]
b^1T_2		4.28	4.09	4.10	4.12	4.60	4.61 [200]
c^1T_2		5.15	4.92	4.94	4.92	4.94	–
r_{Cr-O} (Å)		1.661	1.683	1.681	1.683	1.667	1.660 [200]
RuO_4							
a^1T_2	3.33	3.19	3.04	3.05	3.09	2.93	3.22 [201]
b^1T_2		3.98	3.80	3.82	3.85	3.85	4.00 [201]
c^1T_2		5.25	5.01	5.03	5.05	4.69	4.96 [201]
r_{Ru-O} (Å)		1.712	1.732	1.730	1.732	1.735	1.660–1.850 [202]

[a] Energies calculated at the DFT/LDA/TZP optimized geometries; [b] energies calculated at the DFT/BP/TZP optimized geometries; [c] energies calculated at the DFT/PW91/TZP optimized geometries; [d] energies calculated at the DFT/B3LYP/LAND2DZ optimized geometries

tained for the CrO_4^{2-} complex. The ΔSCF-DFT energy of the lowest dipole allowed $^1T_2$ state is satisfactory in the case of RuO_4, and somewhat overestimated in the case of CrO_4^{2-}.

Very recently the TDDFT approach has been used [99] for the calculation of the vertical excitation energies of an open shell Werner complex, the molybdenum(V) ion, $[MoOCl_4]^-$. The LF as well as LMCT transition energies calculated by TDDFT were in admirable agreement with the experimental electronic absorption and MCD spectra, with a median average deviation (MAD) of 0.13 eV. Among the exchange functionals tested, the hybrid ones, namely B3LYP and B3P86, yielded better results, particularly the B3P86. As in some of the previously discussed complexes, in this case a marked sensitivity of the calculated excitation energies to the ground state geometry was observed, the Mo=O bond distance and the O–Mo–Cl angle being the most critical geometrical parameters.

Another example of the application of TDDFT methods to the electronic spectroscopy of Werner complexes concerns the transition metal hexafluoride series, CrF_6, MoF_6, WF_6. Among these transition metal hexafluorides, CrF_6 is peculiar in that the structure and existence itself of this complex has been debated rather intensively in the literature, both by experimentalists and theoreticians. The synthesis of CrF_6 was first reported in 1963 by Glemser et al [100] and never successfully repeated. However, in 1985 Hope et al [101] prepared a very similar material and also reported a matrix isolation IR spectrum, containing a single band in the stretching region, consistent with an octahedral CrF_6 structure. The available experimental optical spectrum of CrF_6 also relies on matrix isolation experiments by Hope et al [101]. The electronic spectrum of this complex has been investigated theoretically by TDDFT using several exchange-correlation functionals. Adamo and Barone [102], who have computed the vertical excitation energies of the whole MF_6 (M=Cr, Mo, W) series, used the hybrid PBE0 functional; Boulet et al [84] used either asymptotically correct pure functionals, LB94 and LRC, and the hybrid B3LYP functional; Nagle and Baerends [103] used the recent model KS exchange-correlation functional, SAOP. The whole set of TDDFT results for CrF_6 are gathered in Table 11 together with the experimental band maxima reported by Hope et al [101]. The failure of LB94, LRC, and B3LYP functionals to describe the electronic spectrum of this molecule is apparent from the data in Table 11 which shows that all of these functionals dramatically underestimate the excitation energies. This is a real deficiency of these functionals – the poor results are not caused by wrong geometries. By analogy with the slight exaggeration of calculated (BP) and known experimental M–F distances in MoF_6 and WF_6, one might argue that the best optimized BP (largest basis set) Cr–F bond length of 1.745 Å could be slightly overestimated, yielding 1.720±0.004 Å as the best theoretical estimate for the Cr–F bond length [104]. Although a shortening of the Cr–F bond distance from 1.760 to 1.732 Å slightly improves the calculated LB94 excitation energies (see the LB94/LDA results of Table 11), the effect is too small, and also LB94 (and LRC) results at 1.720 Å, although slightly higher and therefore better, would still be much too low. The B3LYP results of Boulet et al [84] exhibit the same severe underestimation of the excitation energies. On the contrary, the PBE0 energies of Adamo and Barone, obtained at the shorter, probably quite accurate, bond distance of 1.718 Å, are much too high. On the basis of these results Boulet et al [84] came to the conclusion that "TDDFT is a disaster for CrF_6". Application of the SAOP approximation to the Kohn-Sham potential, however, which is generally more accurate than LB94, shows that the electronic spectrum of CrF_6 can be adequately described by TDDFT [103]. The SAOP data in Table 11 clearly show that in this case the shape of the functional is by far more important than the ground state geometry for correctly reproducing the electronic spectrum. This is evident when comparing LB94/BP, LB94/PW91, LRC/BP, and SAOP/TZ2P/BP excitation energies, which all refer to the same ground state geometry (r_{Cr-F}=1.760 Å). SAOP results in an extremely large basis (QZ4P, quadruple-zeta with four polarization functions) are also given. As a matter of fact, the larger basis proved to have little effect on the excitation energies if calculated at the same geometry, but it had an indirect effect by shortening the optimized Cr–F bond distance to 1.745 Å. The SAOP results are quite acceptable, very much in line with the best TDDFT results in other cases.

According to the excitation energies reported in Table 12, the performance of the hybrid PBE0 functional improves significantly on going from chromium to the heavier members of the MF_6 series. Indeed, there is an

Table 11 Excitation energies (eV) of CrF_6 obtained via TDDFT methods compared to experimental values

State	LB94/LDA [84][a]	LB94/BP [84][b]	LB94/PW91 [84][c]	LRC/BP [84][b]	B3LYP [84]	PBE0 [102][d]	SAOP/ TZ2P/BP [103][e]	SAOP/ QZ4P/BP [103][f]	Expt [101]
1^1T_{1u}	2.38	2.22	2.22	2.27	2.30	3.76	2.96	3.07	3.31
2^1T_{1u}	3.12	2.96	2.96	2.98	3.21	4.32	3.77	3.87	3.87
3^1T_{1u}	4.21	3.98	3.98	4.04	4.18	5.84	4.68	4.81	4.77
r_{Cr-F} (Å)	1.732	1.760	1.760	1.760	–	1.718	1.760	1.745	–

[a] Energies calculated at the DFT/LDA/TZP optimized geometry; [b] energies calculated at the DFT/BP/TZP optimized geometry; [c] energies calculated at the DFT/PW91/TZP optimized geometry; [d] the excitation energies were calculated at the optimized PBE0 geometry; [e] the excitation energies were calculated at the DFT/BP/TZ2P and DFT/BP/QZ48 optimized geometries (first and second column, respectively)

Table 12 Excitation energies (eV) of MoF_6 and WF_6 computed by TDDFT/ PBE0, compared to SAC-CI and experimental values

State	PBE0 [102]		SAC-CI [105]	Expt [203]
MoF_6				
1^1T_{1u}	5.88[a]	5.98[b]	5.55	5.90
2^1T_{1u}	6.35[a]	6.18[b]	6.61	6.54
3^1T_{1u}	7.35[a]	7.38[b]	7.26	7.12
WF_6				
1^1T_{1u}	6.97[a]	7.13[b]		7.23
2^1T_{1u}	7.85[a]	7.36[b]		8.05
3^1T_{1u}	8.50[a]	8.47[b]		8.60

[a] Energies calculated using the DFT/PBE0 optimized geometries ($r_{Mo\text{-}F}$ = 1.836 Å, $r_{W\text{-}F}$ = 1.839 Å); [b] energies calculated using the experimental geometries from [204] ($r_{Mo\text{-}F}$ = 1.820(3) Å, $r_{W\text{-}F}$ = 1.832(3) Å)

overall reasonable agreement between TDDFT/PBE0 excitation energies and the experimental values, calculated at the experimental geometries. Somewhat improved results are obtained using the PBE0 bond lengths, even if these are not much different. It is remarkable that the small lengthening from R(W–F)=1.832 Å (experimental) to R(W–F)=1.839 Å (PBE0) *increases* the 2^1T_{1u} excitation energy from 7.36 eV to 7.85 eV. As for CrF_6, the geometry dependence found in the literature is (for some excitations) inordinately large. The implication is that only full vibronic analysis, where the vibrational motion of the atoms is accounted for, and full absorption band shapes are calculated, will be able to provide a more definitive assessment of the quality of the TDDFT calculations. In the case of MoF_6 the PBE0 results can be compared to those provided by the much more expensive SAC-CI method [105]. They prove to be reasonably close. We note that the generally better performance of functionals (in this case PBE0) for excitation energies of the second and third row transition metal complexes, is just one example of the general occurrence of more "problems" (both in DFT and ab initio quantum chemical methods) for the first transition metal series, which is caused by the especially tight nature of the 3*d* orbitals. The most difficult case in that respect would be Cu, and an extensive study of Solomon and coworkers [106] has highlighted for the case of $CuCl_4^{2-}$ that indeed the standard functionals yield exceptionally poor spectroscopic properties (bond distances, dissociation energy, spin density, excitation energies) for this molecule, which could be related to wrong positioning of the very tight *d* levels in this case.

To end this section, it is worthwhile mentioning the recent TDDFT study by Autschbach et al [107] on the electronic and circular dichroism (CD) spectra of several chiral Werner complexes, since it represents the first application of TDDFT to the computation of the circular dichroism (CD) spectra of transition metal complexes. The absorption and CD spectra of the

Co^{III} complexes Λ-$[Co(en)_3]^{3+}$, Δ-$[Co(en)_2(acac)]^{2+}$, *cis*-Δ-$[Co(en)_2Cl_2]^{2+}$, *cis*-Δ-$[Co(en)_2(CN)_2]^{+}$, Δ-$[Co(en)_2CO_3]^{+}$, Λ-$[Co(tn)_3]^{3+}$, and for comparison also Λ-$[Rh(en)_3]^{3+}$ (en=ethylenediamine, tn=trimethylenediamine, acac= pentane-2,4-dionato=acetylacetonato) were studied using TDDFT. The resulting simulated CD spectra were generally in good agreement with experimental spectra after corrections for systematic errors in a few of the lowest excitation energies were applied. This allowed for an interpretation and assignment of the spectra for the whole experimentally accessible energy range (UV/vis). Solvent effects on the excitations were estimated via inclusion of a continuum solvent model, the "COnductor-like continuum Solvent MOdel" (COSMO) [108, 109]. This significantly improved the computed excitation energies for CT bands for complexes of charge +3, but had only a small effect on those for neutral or singly charged complexes. The energies of the weak $d \rightarrow d$ transitions of the cobalt complexes were systematically overestimated. This was traced by Autschbach et al [107] to deficiencies in the density functionals (in this study the asymptotically correct SAOP functional and a number of gradient-corrected functionals were tested). The errors were much smaller for the 4*d* metal complex, indicating that 3*d* metals are much more sensitive to approximations in the functionals than 4*d* or 5*d* metals. The same is true for the treatment of correlation effects in wave function-based ab initio methods [110]. Taking these systematic errors and the effect of a solvent into consideration, TDDFT computations have shown to be a reliable tool for the assignment and interpretation of CD spectra of chiral transition metal complexes, as they proved to be for organic systems [111–113].

5 Metal Complexes Containing α-Diimine Ligands

5.1 Metal α-Diimine Complexes

Metal complexes with α-diimine ligands have proven to be very effective in photocatalysis and in photovoltaic conversion [114]. They have also shown good luminescence and electroluminescence abilities, with performances approaching those of more traditional solid state semiconducting devices [115]. Their peculiar molecular properties and the possibilities of related technological applications are directly related to the characteristics of the lowest excited states. The large size of the ligands and the presence in many of these systems of second and third row transition metals have presented challenges to theoretical approaches to the excited states of metal α-diimine complexes. Indeed, until the advent of TDDFT methods, very few accurate theoretical investigations of the excited state properties of metal α-diimine

Table 13 Energies of the absorption maxima of *cis*-[Ru(4,4′-COOH-2,2′-bpy)$_2$(NCS)$_2$] in ethanol and water compared to the corresponding calculated values (energies in eV)

TDDFT						Expt [125, 126]	
Gas-phase			H_2O		C_2H_5OH		
LB94 [121][a]	BPW91 [121][a]	PBE0 [120][b]	BPW91 [121][a]	PBE0 [120][b]	BPW91 [121][a]	H_2O (pH=1)	C_2H_5OH
1.26	1.43	2.24	2.08	2.57	1.97	2.38	2.30
2.50	2.70	3.01	2.92	3.12	2.86	3.18	3.12
			3.69		3.70	3.97	3.95

[a] The excitation energies were calculated at the DFT/BPW91 optimized geometry (C_2 symmetry) of the *cis*-[Ru(4,4′-COOH-2,2′-bpy)$_2$(NCS)$_2$ complex, and solvent effects were evaluated using COSMO [108, 109, 129]; [b] the excitation energies were calculated at the DFT/PBE0 optimized geometry (C_2 symmetry) of the *cis*-[Ru(bpy)$_2$(NCS)$_2$] model complex and solvent effects were evaluated using PCM [128]

complexes had appeared [116]. In the last three years TDDFT studies concerning the excited states of Al(III) [117], Cu(I) [118, 119], Ru(II) [120–122], Os(II) [120], Ir(III) [123], and Pt(II) [124] complexes with a variety of α-diimine ligands have been published. The nature of both the ground and excited states involved in the absorption and/or photoemission, and its tuning by environmental (solvent and/or crystal) effects are the key issues addressed in these studies. An example of the application of TDDFT methods to excited state properties of metal α-diimine complexes is illustrated in Table 13. It concerns the *cis*-[Ru(4,4′-COOH-2,2′-bpy)$_2$(NCS)$_2$], a polypyridyl molecular dye which has been identified as a particularly efficient photosensitizer because of its broad range of visible light absorption, relatively long-lived excited states, and high thermal stability. The visible absorption bands of ruthenium-polypyridyl complexes are usually assigned to MLCT transitions [125], in which an electron is promoted from a Ru-based MO to the empty π^* orbitals of the bipyridine ligands. In particular, the experimental absorption spectrum of the *cis*-[Ru(4,4′-COOH-2,2′-bpy)$_2$(NCS)$_2$] complex in ethanol exhibits two bands in the visible and near-UV regions centered at 538 nm (2.30 eV) and 398 nm (3.12 eV), and a band in the UV region centered at 314 nm (3.95 eV) [125]. The first two bands are assigned to $t_{2g}\rightarrow\pi^*$ transitions while the latter is due to the intraligand $\pi\rightarrow\pi^*$ transitions of the 4,4′-COOH-2,2′-bispyridine (dcbpy) ligands [125]. In addition, this complex shows a pronounced negative solvatochromism, resulting in a blue-shift of the UV-vis absorption bands in water with respect to ethanol [125, 126]. Such solvent-induced shifts are usually interpreted in terms of the different solvation interactions between the polar groups bound to the metal center (SCN^-, CN^- and so on), and mainly depend on solvent polarity and hydrogen bonding [127]. The excited state properties of some M(bpy)$_2$L$_2$ complexes (M=Ru, Os, L=CN, SCN, bpy=2, 2′-bipyridine), both in the gas-phase and in aqueous solution, have been analyzed by Adamo et al [120] at TDDFT/PBE0 level, using the Polarizable Continuum Model (PCM) [128] to take into account solvent effects. More recently, the main features of the near UV-vis spectrum of *cis*-$[Ru(4,4'\text{-}COOH\text{-}2,2'\text{-}bpy)_2(NCS)_2]$ have been investigated at TDDFT level by Fantacci et al [121] both in gas-phase and in two solvents, ethanol and water. In this study, two different exchange-correlation functionals were used, BPW91 and the asymptotically correct LB94. Solvent effects were taken into account using the COnductor-like Screening MOdel (COSMO) [108, 109, 129]. As inferred from the TDDFT results by Fantacci et al [121] gathered in Table 13, inclusion of the solvent effect is essential to correctly reproduce the energy and shape of the experimental spectrum. Indeed, while the computed spectrum in vacuo deviates from the experimental one both in energy and number of bands, the spectra calculated in the presence of the solvent are in good agreement with experiment, even though both spectra are red-shifted by ~0.3 eV with respect to the experiment. The experimentally observed blue-shift of the spectrum in water with respect to ethanol is nicely reproduced theoretically. In agreement with previous assignment [125], TDDFT calculations predict an intraligand $\pi \rightarrow \pi^*$ character for the UV band. The first two absorption bands are found however to originate from mixed ruthenium-NCS to bipyridine-π^* transitions rather than to pure MLCT transitions.

As for the PBE0 results, the excitation energies calculated for the two lowest excited states of the model complex, *cis*-$[Ru(bpy)_2(NCS)_2]$, change very little when the solvent effects are included. The excitation energies computed in vacuo already reproduce the two lowest bands observed in the spectrum of the *cis*-$[Ru(4,4'\text{-}COOH\text{-}2,2'\text{-}bpy)_2(NCS)_2]$ complex in water quite well.

5.2
Mixed-Ligand Metal α-Diimine Carbonyls

Among the metal complexes containing α-diimine ligands, mixed-ligand metal α-diimine carbonyls are of particular interest in view of their unconventional photochemical, photophysical, and electrochemical properties. Variation of the degree of electronic delocalization over the α-diimine group, of the metal center, and the other ligands allows us to modify their spectroscopic and photochemical behavior. With the aim of comparing CASSCF/CASPT2 and TDDFT/B3LYP theoretical approaches and their applicability to the assignment of electronic absorption spectra of organometallic compounds, Turki et al [191] carried out a combined CASPT2 and TDDFT study of the UV-visible spectra of $[Ru(E)(E')(CO)_2(iPr\text{-}DAB)]$ (E=$SnPh_3$, Cl; E′=$SnPh_3$, Cl, CH_3; iPr-DAB=*N,N′*-di-isopropyl-1,4-diaza-1,3-butadiene). These complexes, which represent an important class of low-valent metal complexes containing electron-acceptor ligands, π donors such as halides,

Table 14 CASSCF/CASPT2 and TDDFT/B3LYP excitation energies (eV), oscillator strengths f (in parentheses), and assignments of the low-lying electronic transitions of $[Ru(SnH_3)(CH_3)(CO)_2(Me\text{-}DAB)]$ and $[Ru(Cl)(CH_3)(CO)_2(Me\text{-}DAB)]$ [2, 191]

State	CASSCF/CASPT2[a]		TDDFT/B3LYP[a]		Expt [191]
	Character	Energy (f)	Character	Energy (f)	
$[Ru(SnH_3)(CH_3)(CO)_2(Me\text{-}DAB)]$					
b^1A'	$\sigma_{Me\text{-}Ru\text{-}Sn}\rightarrow\pi^*_{DAB}$	2.55 (0.11)	$\sigma_{Me\text{-}Ru\text{-}Sn}\rightarrow\pi^*_{DAB}$	2.69 (0.06)	2.32
a^1A''	$4d_{xy}\rightarrow\pi^*_{DAB}$	2.60 (0.004)	$4d_{xy}\rightarrow\pi^*_{DAB}$	2.71 (0.002)	2.78
c^1A'	$4d_{xz}\rightarrow\pi^*_{DAB}$	3.21 (0.22)	$4d_{xz}\rightarrow\pi^*_{DAB}$	3.42 (0.067)	3.17
$[Ru(Cl)(CH_3)(CO)_2(Me\text{-}DAB)]$					
a^1A''	$4d_{xy}\rightarrow\pi^*_{DAB}$	2.24 (0.005)	$p_{Cl}/4d_{xy}\rightarrow\pi^*_{DAB}$	1.81 (0.000)	2.17
b^1A'	$4d_{xz}\rightarrow\pi^*_{DAB}$	2.83 (0.23)	$p_{Cl}/4d_{xz}\rightarrow\pi^*_{DAB}$	2.03 (0.011)	2.72
c^1A'	$\sigma_{Me\text{-}Ru\text{-}Cl}\rightarrow\pi^*_{DAB}$	3.35 (0.038)	$\sigma_{Me\text{-}Ru\text{-}Cl}\rightarrow\pi^*_{DAB}$	2.91 (0.015)	3.51

[a] Energies and oscillator strengths calculated at the DFT/B3LYP optimized geometries (C_s symmetry)

and/or σ-bonded alkyl or metal fragment ligands simultaneously, have great potential as luminophores, photosensitizers, and visible light photoinitiators of radical formation. With respect to their spectroscopic and photochemical behavior, they can be classified into two distinct groups: (1) complexes in which at least one of the E, E′ ligands is a halide; (2) complexes where E, E′ are either an alkyl group or metal fragments. In Table 14 the TDDFT and CASSCF/CASPT2 transition energies calculated for $[Ru(SnH_3)(CH_3)(CO)_2(Me\text{-}DAB)]$ and $[Ru(Cl)(CH_3)(CO)_2(Me\text{-}DAB)]$, taken as prototypes of the two distinct groups above mentioned, are reported together with the main features of the experimental absorption spectra. As for the non-halide complex, TDDFT calculations reproduce the visible and near-UV region of the spectrum of $[Ru(SnPh_3)(CH_3)(CO)_2(Me\text{-}DAB)]$ well, which is characterized by a strong absorption at around 2.3 eV, a weak shoulder at ~2.8 eV, and a weak but distinct band at 3.17 eV. As illustrated in Table 14, a very good agreement exists between TDDFT and CASSCF/CASPT2 concerning the energy and character of the bands. Notably, both theoretical approaches, in agreement with the experimental assignment [130, 131], attribute the lowest strong visible absorption band to a SBLCT (sigma-bond-to-ligand-charge transfer) transition, the $\sigma_{Me\text{-}Ru\text{-}Sn}\rightarrow\pi^*_{DAB}$, and the near-UV features to $4d_{Ru}\rightarrow\pi^*_{DAB}$ MLCT transitions.

Unlike in the case of non-halide complexes, TDDFT and CASSCF/CASPT2 techniques lead to different results for the halide substituted molecules. As inferred from Table 14, where the representative case of $[Ru(Cl)(CH_3)(CO)_2(Me\text{-}DAB)]$ is illustrated, TDDFT systematically underestimates the transition energies of the halide complexes, although it reproduces the general spectral pattern well and also the increase in molar absorptivity of the main visible absorption band on going from the halide to the non-halide complexes. The main visible absorption band of the halide complexes [Ru(Cl)(Me)$(CO)_2$(iPr-DAB)] and [Ru$(Cl)_2(CO)_2$(iPr-DAB)] occurs at higher energy than in the non-halide molecules, at 459 nm (2.76 eV) and 469 nm (2.64 eV) respectively. It is ~3 and 6 times weaker than the visible band observed for [Ru($SnPh_3$)(Me)$(CO)_2$(iPr-DAB)] and [Ru$(SnPh_3)_2(CO)_2$(iPr-DAB)]. Noteworthy dramatic differences were found between TDDFT and CASSCF/CASPT2 in the description of the two lowest excited states. The TDDFT approach predicts for these states a mixed XLCT/MLCT (XLCT stands for a Cl→DAB excitation) character with the XLCT component being predominant, whereas the CASSCF/CASPT2 assigns them to MLCT transitions. This discrepancy stems from a different description of the highest occupied MOs, that are predicted to be $4d_{\mathrm{Ru}}$ by the ab initio approach and p_{Cl} by DFT. The poor performance of TDDFT in predicting the excitation energies of halide containing systems was shown to be substantially invariant to the choice of the exchange-correlation functional and the basis set. One may wonder if one is dealing here with the well-known failure of TDDFT in predicting the excitation energies of charge-transfer states correctly. As recently demonstrated by Drew et al [132], this failure is due to the local character of the approximate xc functionals presently available. Indeed, the population analysis of the excited states of [Ru(Cl)$(CH_3)(CO)_2$(Me-DAB)] and [Ru$(Cl)_2(CO)_2$(Me-DAB)] reveals that the electronic transitions responsible for the CT states we are dealing with are accompanied by a considerable charge transfer from the halide to the α-diimine ligand.

It should be noted that the remarkable charge-transfer observed in these systems does not occur so frequently in the so-called CT states of transition metal complexes, be they LMCT, MLCT or LLCT. Indeed, in most of the cases no net charge-transfer occurs, only a reorganization of the electronic density within the metal or ligand orbitals, as demonstrated long ago by Ziegler et al [94] for the CT states of metal oxides. The classification in terms of LMCT, MLCT, LLCT is based on the character of the orbitals involved in the transitions. However, for instance, excitation of an electron out of a (mostly) ligand orbital to a (mostly) *d* orbital is usually accompanied by a change in the composition of other orbitals (the "passive" orbitals) in such a way that the total *d* population remains practically the same. This is a consequence of the strong *d*-*d* repulsion. Exciting an electron from a ligand to the *d* shell in a one-electron transition will increase the *d* population. However, this increased *d* population will destabilize the *d* levels, with the effect that the *d* mixing into other occupied ligand orbitals diminishes, which has the effect of decreasing the total *d* population. The net effect is usually very little change in the *d* population. This explains why, for the majority of the systems discussed here, the CT states do not suffer the TDDFT failure mentioned above.

6
Metallotetrapyrroles

For the description of the linear and nonlinear optical properties of metallotetrapyrroles, TDDFT methods have proven [133–148] to be an excellent alternative to conventional highly correlated ab initio methods, such as SAC-CI, STEOM-CC, and CASPT2, for which these systems still represent a severe computational challenge, especially when transition metals, lanthanides or actinides are involved. The few highly correlated ab initio calculations dealing with the excited state properties of metallotetrapyrroles that have appeared to date only concern magnesium and zinc porphyrins and porphyrazines [149–151]. Application of TDDFT methods to the electronic spectroscopy of a variety of metallotetrapyrroles, including homoleptic and heteroleptic sandwiches, will be illustrated in this section.

6.1
Porphyrins

Porphyrins and their derivatives play an important role in the biochemistry of all living systems [152]. They form the backbone of pigments such as chlorophyll and heme, which have a universal biological distribution. Therefore, knowledge of these systems and their excited states is essential in understanding a wide variety of biological processes, including oxygen binding, electron transfer, catalysis, and the initial photochemical step in photosynthesis.

The electronic absorption spectra of metalloporphyrins are characterized by very weak absorption in the visible region showing vibrational structure, the so called Q band, an intense feature in the near-UV, the most intense in the spectra, known as the B or Soret band, and, at the higher energy side of the B band, N and M bands, both attributed to the porphyrin ring [153]. Additional, metal dependent L and "extra" bands are also present in the UV region of the spectra of transition metal porphyrins [153]. The origin and the intensities of the lowest absorptions of metal-porphyrins, the Q and the B bands, were successfully explained by Gouterman's four-orbital model [154–156]. Historically, the cyclic polyene model for the porphyrinic skeleton has played a large role [155, 157, 158] and is still often cited as providing a qualitative basis for the understanding of the weak Q and strong B band system of porphyrins. The LCAO-MO treatment that gives rise to the four orbital model improves upon this by explicitly taking into account the structure of the porphyrin ring [156]. According to the four-orbital model, the B and Q bands can be described in terms of transitions between a pair of top filled (a_{1u} and a_{2u}) and lowest empty (the doubly degenerate e_g) π orals of the porphyrin macrocycle. The near degeneracy of the $a_{1u}{}^1e_g{}^1$ and $a_{2u}{}^1e_g{}^1$ excited configurations leads to a strong configuration interaction that results in a high-lying state corresponding to the B band, and a low-lying state corresponding to the Q band. The configuration mixing combines the transition dipoles of the individual one-electron transitions in such a way that the B band contains nearly all of the intensity, while the Q band is weak. The accuracy of TDDFT methods and the impact of different functionals in predicting the important features of the optical spectra of metalloporphyrins such as the relative intensities of the Q and B band systems, the nature of the higher energy additional bands, as well as the presence of the metal *d* states, will be illustrated by two examples: the first concerns unligated porphyrin complexes (MgP, ZnP, and NiP); the second concerns an iron porphyrin with an imidazole and CO ligand at the iron, modeling the active centers of hemoproteins.

The TDDFT excitation energies and oscillator strengths calculated for the lowest allowed ^1E_u excited states of MgP [138], ZnP [138, 140], and NiP [137, 138] complexes in their optimized D_{4h} geometry, are reported in Table 15 and compared to CASPT2 [150] and Multireference Moller-Plesset perturbation (MRMP) theory [151] results.

Considering first NiP, the two lowest ^1E_u states computed at TDDFT/SAOP level at 2.40 eV and 3.23 eV account for the Q and B bands in the spectra very well. As a consequence of the quasi degeneracy of the a_{1u} and a_{2u} Gouterman orbitals, these states are a nearly 50–50 mixture of the $a_{1u}{\rightarrow}e_g{}^*$ and $a_{2u}{\rightarrow}e_g{}^*$ transitions. The 1^1E_u state has a very small oscillator strength (0.0052), due to opposite directions of the two large transition dipoles, while the 2^1E_u where these transition dipoles are parallel has a large oscillator strength (1.0214). The 3^1E_u state calculated at 3.45 eV with very small oscillator strength has a mono-transitional character consisting entirely of the $2b_{2u}{\rightarrow}e_g{}^*$, $\pi{\rightarrow}\pi^*$ transition. The 4^1E_u and 5^1E_u states calculated at 3.77 eV and 4.00 eV respectively nicely account for the N and L bands, with maxima at 3.70 eV and 4.20 eV in the gas-phase absorption spectrum of NiOEP (OEP=octaethylporphyrin). These states have a mixed $\pi{\rightarrow}\pi^*$ ($3a_{2u}{\rightarrow}5e_g$)/ $d_\pi{\rightarrow}\pi^*$ ($4e_g{\rightarrow}2b_{1u}$) character. They consist of the same transitions with approximately reversed weights, the $\pi{\rightarrow}\pi^*$ character being predominant in the 4^1E_u, the $d_\pi \rightarrow \pi*$ being predominant in the 5^1E_u. The TDDFT/SAOP description of the N and L bands is consistent with the experimental observations. Indeed, the N band is quite constant in the vapor-phase spectra of metal-porphyrins, whereas the L band, which is observed only in porphyrin complexes of transition metals with incomplete *d* shells, varies considerably with the metal in both energy and intensity [153].

At BP level, the energy, intensity and composition of the Q state (1^1E_u) do not change. It is not so however for the higher energy excited states that are calculated systematically at lower energy than at SAOP level, leading in the case of the N (4^1E_u) and L (5^1E_u) states to a large discrepancy (6% and 9% respectively) with respect to experiment. It is worth mentioning that at BP level the B state has a non-negligible $d_\pi{\rightarrow}\pi^*$ MLCT character, and the

Table 15 Excitation energies (eV) and oscillator strengths (in parentheses) of the lowest allowed excited states of MgP, ZnP, and NiP (calculated by TDDFT methods), compared to available CASPT2 and MRMP values, and to experimental data

State	TDDFT						CASPT2[d]		MRMP[e]		Expt
	BP[a]		SAOP[b]		B3LYP[c]						
MgP											
1^1E_u	2.24 (0.0006)	Q	2.23 (0.0004)	Q			1.66 (0.004)	Q	2.00 (0.0106)	Q	2.14[f], 2.2[g] (Q)
2^1E_u	3.16 (0.1439)		3.21 (0.3436)				2.66 (0.824)	B	3.07 (1.5613)	B	3.18[f] (B)
3^1E_u	3.27 (1.0692)	B	3.29 (1.0080)	B			3.11 (0.150)	sh			
4^1E_u	3.63 (0.8476)	N	3.64 (0.4140)	N			3.42 (0.569)	N			3.81[f] (N)
ZnP											
1^1E_u	2.29 (0.0005)	Q	2.28 (0.0019)	Q	2.44 (0.002)	Q			2.11 (0.0121)	Q	2.21[h], 2.18[i], 2.16[j] (Q)
2^1E_u	3.20 (0.2218)		3.24 (0.4452)		3.54 (0.889)	B			3.21 (1.6281)	B	3.09[h], 3.13[i], 3.19[j] (B)
3^1E_u	3.31 (1.1354)	B	3.32 (1.0062)	B	3.84 (0.052)						3.3[h]
4^1E_u	3.65 (0.5344)	N	3.66 (0.5198)	N	4.29 (0.175)	N					4.07[i], 3.81[j] (N)
NiP											
1^1E_u	2.40 (0.0048)	Q	2.40 (0.0052)	Q							2.28[k], 2.22[l] (Q)
2^1E_u	3.15 (0.6068)	B	3.23 (1.0214)	B							3.11[k], 3.22[l] (B)
3^1E_u	3.36 (0.1376)		3.45 (0.0009)								
4^1E_u	3.48 (0.3424)	N	3.77 (0.0294)	N							3.70[l] (N)
5^1E_u	3.81 (0.2330)	L	4.00 (0.2526)	L							4.20[l] (L)

[a] See [205] for MgP, [137] for ZnP and NiP; [b] see [138]; [c] see [140]; [d] see [150]; [e] see [151]; [f] gas-phase spectrum of Mg-Etio, from [153]; [g] low-temperature matrix experiment on MgP, from [206]; [h] data for ZnP in C_6H_6, from [161]; [i] data for ZnP in MeOH from [159]; [j] gas-phase spectrum of ZnOEP from [153]; [k] CS_2 solution spectrum of NiP from [207]; [l] gas-phase spectrum of NiOEP from [153]

N and L states are predicted to have reverse character, $d_\pi \rightarrow \pi^*$ the former, $\pi \rightarrow \pi^*$ the latter, indicating a shift to lower energy of the MLCT states on going from SAOP to BP functional. Since SAOP results are more in line with the experiment than BP results, both in energy and character, it can be argued that the MLCT states are better described by the SAOP functional.

As inferred from Table 15, in the case of MgP and ZnP where the excited states invariably have $\pi \rightarrow \pi^*$ character, since MLCT transitions are out of business, SAOP and BP functionals provide nearly the same results. In MgP and ZnP, SAOP and BP calculations put the weak intensity $b_{2u} \rightarrow e_g^*$ transition mostly in the 2^1E_u excited state, the Gouterman transitions entering with large weight in the much more intense and nearly degenerate 3^1E_u. On account of some Gouterman transition character, the 2^1E_u state of MgP and ZnP however has larger intensity than the related NiP 3^1E_u. These two states were assigned by Baerends et al [138] to the B band. This band is observed at 3.18 eV in the gas-phase spectrum of Mg-Etio [153], at 3.19 eV in the gas-phase spectrum of ZnOEP [153], and in the range 2.95–3.18 eV in the solution spectra of ZnP [159–162].

At variance with TDDFT/SAOP and TDDFT/BP, MRMP calculations on MgP and ZnP [151], CASPT2 calculations on MgP [150] and TDDFT/B3LYP calculations on ZnP [140] put the $a_{1u} \rightarrow e_g^*$ and $a_{2u} \rightarrow e_g^*$ Gouterman transitions in the intense 2^1E_u state. The $b_{2u} \rightarrow e_g^*$ transition is predicted by CASPT2 and TDDFT/B3LYP mostly in the weak 3^1E_u state computed at higher energy than the 2^1E_u (0.45 eV in the CASPT2 calculations on MgP and 0.3 eV in the TDDFT/B3LYP calculations on ZnP) and consequently they assign the 2^1E_u state to the B band and the 3^1E_u state to a shoulder of the B band. However, the position of the B band is not well represented by these calculations. From the data of Table 15 it is apparent that CASPT2 calculations underestimate the B band of MgP by ~0.5 eV and that TDDFT/B3LYP calculations overestimate the B band of ZnP by ~0.4 eV. The MRMP and both SAOP and BP results are much more in line with the experiment as regards the B band position.

TDDFT also provides a satisfactory assignment of the UV region of the electronic spectra of MgP, ZnP, and NiP, although some ambiguities still hold, mainly because of the broadening of the bands in this region.

The second example of the application of TDDFT to the electronic spectroscopy of metalloporphyrins concerns a CO-ligated iron porphyrin, a system that models the active centers of hemoproteins, recently investigated by Head-Gordon et al [146, 147] in the context of a theoretical study of the initial step of the photodissociation pathway of CO-ligated heme.

In view of the potential application of TDDFT methods to the spectroscopy of bioinorganic systems, the dependence of the excitation energies on the exchange-correlation functional and the basis set was carefully tested. In Table 16, the TDDFT results are compared to the known experimental data for the Q and B bands of CO-ligated mioglobin (Mb) [163], the SAC-CI calcula-

Table 16 The TDDFT energies of the Q and B bands of CO-ligated heme calculated with different exchange-correlation functional and basis sets, compared to CIS, SAC-CI and experimental data

Excited state	Q_1	Q_2	B_1	B_2
SVWN [147][a]				
LANL2DZ	1.5150	1.5369	3.0093	3.0204
6-31G	1.3338	1.3568	2.8994	2.9215
6-31G*	1.3908	1.4128	2.9038	2.9250
BLYP [147][a]				
LANL2DZ	1.8226	1.8428	3.0493	3.0559
6-31G	1.5057	1.5271	2.9292	2.9524
6-31G*	1.5727	1.5933	2.9266	2.9434
B3LYP [147][a]				
LANL2DZ	2.4039	2.4177	3.3458	3.3591
6-31G	2.3354	2.3961	3.6902	3.6976
6-31G*	3.4027	2.4127	3.3140	3.3250
CIS [147][a]				
LANL2DZ	2.5620	2.5633	4.4428	4.4820
6-31G*	2.4898	2.4907	4.4231	4.4629
SAC-CI [164]				
	1.84	1.94	3.36	3.41
Expt[b]				
	2.18	2.30	2.96	3.16

[a] The energies were calculated at the DFT/B3LYP/LANL2DZ optimized geometry (C_s symmetry) of the model complex; [b] spectrum of CO ligated myoglobin, from [163]

tions by Tokita and Nakatsuji [164], and the configuration interaction singles (CIS) values [147]. The SVWN and BLYP functionals give values for the B band that are very close to the experimental value, whereas the Q bands are underestimated by up to 0.8 eV, suggesting an unbalanced description of the Q and B states by these functionals. Only the B3LYP functional shows a consistent error for the Q and B states and overestimates them by 0.2–0.3 eV. Furthermore, the calculated excitation energies depend only slightly on the choice of the basis set. Neither the use of an effective core potential, nor the addition of polarization functions has a significant influence on the results for the excitation energies. As inferred from Table 16, CIS overestimates the excitation energies drastically, and SAC-CI inconsistently treats the Q and B bands.

Altogether, the TDDFT results indicate a large spread of the calculated energy values between different functionals. On going from the pure SVWN and BLYP functionals to the hybrid B3LYP there is really a jump in the energy values, particularly for the Q bands. SVWN and BLYP results are not that good, at least not of the same quality as SAOP in the previous cases.

6.2
Porphyrazines, Benzoporphyrins, and Phthalocyanines

The effects on the optical spectra of introducing azabridges, benzo rings, and both aza bridges and benzo rings into the basic porphyrinic ring, yielding the most representative tetrapyrrole ligands – porphyrazines (Pz), tetrabenzoporphyrins (TBP), and phathalocyanines (Pc) – have been successfully interpreted by TDDFT methods. Several TDDFT studies have appeared concerning the excited states of transition and non transition metal complexes of porphyrazines [137, 138, 140, 142, 144], tetrabenzoporphyrins [137, 140, 142, 144], and phthalocyanines [135, 137, 140, 142, 144, 148]. Two complete studies are especially worth mentioning, concerning the electronic spectroscopy of the nickel tetrapyrrole series [137], NiP, NiPz, NiTBP, and NiPc, and the zinc homologue [140] where the main spectral changes were interpreted on the basis of electronic structure changes upon variation of the macrocycle framework. According to the analysis of the Kohn-Sham MOs performed in these studies, the tetrabenzo annulation of the porphyrin ring causes a strong destabilization of the a_{1u} orbital (the HOMO in NiP, the HOMO-1 in ZnP), which is no longer (nearly)-degenerate with the a_{2u}, and introduces additional low-lying π^* levels into the virtual spectrum. *Meso*-tetraaza substitution stabilizes all levels, but specially the a_{2u}, which has large amplitude on the aza bridges, to the effect that the porphyrin a_{1u}/a_{2u} degeneracy is removed. The aza bridges also introduce additional N_b (bridging nitrogen)-lone pair levels into the valence region. In phthalocyanines, the effects of aza bridges and benzo rings on the a_{1u} and a_{2u} orbitals are both operative. The a_{1u} is destabilized by the benzo rings, although it lies at somewhat lower energy than in TBPs due to the general downward shift of the Pc levels induced by the aza bridges, while the a_{2u} is shifted very far down with respect to the a_{1u} due to the combined effect of the upshift of the a_{1u} caused by the antibonding with the benzo rings and the downshifting effect of the aza bridges on the a_{2u}. Actually, a thorough electronic structure analysis of ZnPc [135] and NiPc [137] has revealed that of the two highest occupied a_{2u} orbitals, it is the lower lying, the $4a_{2u}$, that should be identified as the Gouterman orbital.

As inferred from the data in Table 17 where the excitation energies and oscillator strengths calculated for MgPz, ZnPz, and NiPz are reported and compared to experimental data, the main features of the spectra of metalloporphyrazines are quite nicely reproduced by TDDFT calculations. The lifting of the (near-)degeneracy of the occupied a_{1u} and a_{2u} Gouterman orbitals in MPz causes the $a_{1u} \rightarrow e_g^*$ and $a_{2u} \rightarrow e_g^*$ transitions to mix very little. This has important consequences for the nature and intensity of the lowest excited states. The 1^1E_u excited state is in all porphyrazines mainly described by the $a_{1u} \rightarrow e_g^*$ transition, the $a_{2u} \rightarrow e_g^*$ entering with only a minor weight. The cancellation of the transition dipoles of the Gouterman configurations that

Table 17 Excitation energies (eV) and oscillator strengths (in parenthesis) of the lowest allowed excited states of MgPz, ZnPz, and NiPz, as calculated by TDDFT methods, compared to experimental data

TDDFT							Expt
State	BP [138][a]		SAOP [137][b]		B3LYP [140][c]		
MgPz							
1^1E_u	2.33 (0.1722)	Q	2.34 (0.2308)	Q			2.08[d] (Q)
2^1E_u	2.75 (0.0244)		2.81 (0.0235)				
3^1E_u	3.05 (0.0655)	B	3.08 (0.0540)	B			3.31[d] (B)
4^1E_u	3.66 (1.6092)	N	3.74 (1.7084)	N			3.61[d] (N)
ZnPz							
1^1E_u	2.37 (0.2052)	Q	2.34 (0.2530)	Q	2.48 (0.148)	Q	2.13[e] (Q)
2^1E_u	2.80 (0.0238)		2.81 (0.0238)		3.38 (0.018)	$\pi\rightarrow\pi^*$	
3^1E_u	3.17 (0.0305)	B	3.16 (0.0135)	B	3.78 (0.114)	B_1	3.44[e] (B)
4^1E_u	3.66 (1.4176)	N	3.69 (1.4560)	N	4.04 (0.714)	B_2	3.73[e] (N)
NiPz							
1^1E_u	2.43 (0.2234)	Q	2.42 (0.2692)	Q			2.11[f] (Q)
2^1E_u	3.00 (0.0222)	sh	3.00 (0.0226)	sh			3.30[f] (sh)
3^1E_u	3.16 (0.0008)		3.38 (~0)				
4^1E_u	3.42 (0.0064)		3.51 (0.1465)	B			3.65[f] (B)
5^1E_u	3.54 (0.0940)	B	3.67 (0.0300)				
6^1E_u	3.78 (0.9382)	N	3.89 (0.9220)	N			3.91[f] (N)

[a] The excitation energies and oscillator strengths were calculated at the DFT/BP/TZP D_{4h} optimized geometry; [b] the excitation energies and oscillator strengths were calculated at the DFT/BP/TZP D_{4h} optimized geometry; [c] the excitation energies and oscillator strengths were calculated at the DFT/B3LYP/6-31G(d) D_{4h} optimized geometry; [d] data for MgOEPz (OEPz is octaethylporphyrazine) in *n*-hexane from [137]; [e]data for ZnP from [208]; [f] data for NiOEPz in *n*-hexane from [137]

occurs in the lowest excited state of porphyrins, leading to very low intensity of the Q bands, occurs to a much lesser extent in MPz, leading, in agreement with experiment, to an intense Q band. The observed red shift of the Q band is also quite nicely reproduced, although the energy of the Q state is slightly overestimated by TDDFT.

As for the near-UV region of the electronic spectra of porphyrazines, it is characterized by a broad and structured band containing a main peak denoted according to the nomenclature used by Weiss et al [165] as N, and a pronounced shoulder to the red side of the N band, denoted as B. The energies of the B and N bands as well as the red shift of the B–N system going from the transition to the non transition metal porphyrazines are quite well reproduced by both BP and SAOP functionals [137, 138]. TDDFT/B3LYP calculations on ZnPz by Nguyen and Pachter [140] provide substantially the same qualitative interpretation of the optical spectrum as TDDFT/SAOP and TDDFT/BP, apart from the use of a different notation for the UV bands, which are denoted by these authors as B_1 and B_2 (in analogy with phthalocyanines, vide infra), rather than B and N. However, the TDDFT/B3LYP exci-

Table 18 Excitation energies (eV) and oscillator strengths (in parentheses) of the lowest allowed ^1E_u excited states of ZnTBP and NiTBP, as calculated by TDDFT methods, compared to experimental data

TDDFT							Expt
State	BP [137][a]		SAOP [137][b]		B3LYP [140][c]		
ZnTBP							
1^1E_u					2.18 (0.173)	Q	1.98 (0.30)[d] (Q)
2^1E_u					3.28 (1.121)	B	3.06 (1.60)[d] (B)
NiTBP							
1^1E_u	2.08 (0.3070)	Q	2.08 (0.3130)	Q			2.01[e] (Q)
2^1E_u	2.94 (0.7198)	B	2.95 (0.7144)	B			3.00[e] (B)

[a] The excitation energies and oscillator strengths were calculated at the DFT/BP/TZP D_{4h} optimized geometry; [b] the excitation energies and oscillator strengths were calculated at the DFT/BP/TZP D_{4h} optimized geometry; [c] the excitation energies and oscillator strengths were calculated at the DFT/B3LYP/6–31G(d) D_{4h} optimized geometry; [d] gas-phase spectrum of ZnTBP from [166], the absolute oscillator strength values reported in parenthesis were obtained by assuming that the values for the Q and B bands measured in solution hold in vapor; [e] data for NiTBP in dichloromethane from [137]

tation energies are too high compared to the experimental values, which confirms the tendency of the B3LYP functional to overestimate the excitation energies of tetrapyrrole complexes. In the case of NiPz, where the B and N states involve MLCT ($d_\pi \rightarrow \pi*$) and LMCT ($n \rightarrow d_{\sigma*}$) transitions to a large extent, BP and SAOP results are, for the reasons discussed above, somewhat different; SAOP excitation energies being, as in the case of NiP, closer to the experimental values.

We can infer from the data in Table 18 that the changes in energy and intensity of the Q and B band observed upon tetrabenzo annulation of the porphyrin ring, are very nicely reproduced by TDDFT calculations. B3LYP results on ZnTBP [140], and both BP and SAOP results on NiTBP [137] come to an unambiguous assignment of the 1^1E_u state to the rather intense Q band, and of the 2^1E_u state to the narrow, very intense, Soret (B) band. Due to the upshift of the a_{1u} Gouterman orbital upon tetrabenzo annulation, the Q state (1^1E_u) is mainly described by the $a_{1u} \rightarrow e_g{}^*$ transition, the $a_{2u} \rightarrow e_g{}^*$ entering with only a minor weight. The oscillator strength associated with the 1^1E_u state is quite large for the same reason as in metalloporphyrazines, in line with the Q band being much more intense in tetrabenzoporphyrins and porphyrazines than in porphyrins. The upshift of the a_{1u} also causes the Q state in ZnTBP and NiTBP to lie at lower energy than in the porphyrin analogs, which fits in with the observed red shift of the Q band on going from porphyrins to tetrabenzoporphyrins. In the region to the blue of the B band, the optical spectra of ZnTBP and NiTBP show significant differences [137, 166], MLCT and LMCT states coming into play in the case of the nickel complex. These differences are also satisfactorily accounted for by TDDFT calculations [137].

As for metallophthalocyanines, a considerable effort has been devoted to the characterization of the excited states of these tetrapyrroles, their derivatives, and assemblies thereof by a variety of optical spectrometric techniques including circular dichroism (CD), magnetic circular dichroism (MCD), and fluorescence spectroscopy, and, in recent years, time-resolved optical spectroscopy. This is not surprising, since metallophthalocyanines and derivatives display interesting properties such as light stability and efficient light absorption in the red and visible region of the spectrum that make them potential candidates for applications in optoelectronics, photoconducting materials [167–169] and photosensitizers for photodynamic therapy (PDT) [170, 171]. Among these metallophthalocyanines, ZnPc has received special attention, mainly because it is ideally suited for the characterization of the excited states of the Pc ring. Due to the d^{10} configuration of the central Zn^{2+} ion, the optical spectra of ZnPc are not complicated by the "additional" MLCT and LMCT bands appearing in the spectra of other transition metal phthalocyanines, yet the D_{4h} symmetry of MPc is retained.

To date, a number of experimental studies of the excited states of ZnPc have been made available, including gas phase [172], thin film [173], solution [174, 175], and argon matrix [176] absorption spectra, solution [174, 175] and argon matrix [176] MCD spectra, solution luminescence spectra [177], and luminescence and magnetic circularly polarized luminescence (MCPL) spectra in an argon matrix [177].

The first investigation into the excited states of ZnPc based on first-principles methods is the TDDFT/SAOP study by Ricciardi et al [135], where the UV-vis and the vacuum ultraviolet region of the electronic spectrum of ZnPc are described in detail. Subsequently, Nguyen and Pachter, in the context of a TDDFT/B3LYP study of the electronic spectroscopy of the zinc tetrapyrrole series [140], ZnP, ZnPz, ZnTBP, and ZnPc, came to a somewhat different interpretation of the Uv-vis spectrum of this phthalocyanine.

In Table 19 the energies and oscillator strengths of the dipole allowed ^1E_u and $^1A_{2u}$ excited states of ZnPc contributing to the Q and B band region, computed at SAOP and B3LYP level, are compared to the experimental data.

Both SAOP and B3LYP assign the 1^1E_u state computed at 1.96 eV and 2.09 eV respectively to the intense Q band located at 1.88 eV in the gas-phase and Ar/matrix spectra, and at 1.85 eV in both room temperature and cryogenic temperature vitrified solution spectra of ZnPc. The (near)-degeneracy of the a_{1u} and a_{2u} Gouterman orbitals which exists in porphyrins is lifted by the electronic effects of the aza bridges and benzo rings, so the mixing of the $(a_{1u}e_g{}^*)$ and $(a_{2u}e_g{}^*)$ configurations and the consequent cancellation of the associated transition dipole moments occurring in the Q band of porphyrins does not occur in phthalocyanines. According to TDDFT calculations, the 1^1E_u is a nearly pure $2a_{1u}\rightarrow 7e_g$ state and its oscillator strength is

Table 19 Excitation energies (eV) and oscillator strengths (in parentheses) of the lowest allowed excited states of ZnPc, as calculated by TDDFT methods, compared to experimental data

TDDFT					Expt			
State	SAOP [135][a]		B3LYP [140][b]		Gas-phase[c]	Solution[d]	solution[e]	Ar matrix[f]
1^1E_u	1.96 (0.7356)	Q	2.09 (0.418)	Q	1.88 (Q)	1.85 (Q)	1.85 (Q)	1.85 (Q)
2^1E_u	2.87 (0.0307)	$\pi \rightarrow \pi^*$	3.37 (0.008)	$\pi \rightarrow \pi^*$				
3^1E_u	3.07 (0.0518)		3.66 (0.172)	B_2			**3.05** (B_1)	3.05 (B_1 origin)
4^1E_u	3.14 (0.3032)	B_1	3.74 (0.328)	B_1		3.21 (B_1)	3.17	
1^1A_{2u}	3.28 (0.0011)	$n \rightarrow \pi^*$	3.93 (0.003)	$B_3(n \rightarrow \pi^*)$				
5^1E_u	3.34 (0.0446)		3.87 (0.337)	$\pi \rightarrow \pi^*$			**3.30** (B_1)	
6^1E_u	3.50 (0.6572)		3.96 (0.116)	$\pi \rightarrow \pi^*$			**3.56** (B_2)	3.71 (B_2)
7^1E_u	3.81 (1.1548)	B_2	4.34 (0.193)	$N(\pi \rightarrow \pi^*)$	3.80 (B)	3.75 (B_2)	**3.81** (B_2)	3.74 (B_1 max)
2^1A_{2u}	4.23 (0.0012)	$B_3(n \rightarrow \pi^*)$					4.05	3.99 (B_3)
14^1E_u	4.77 (0.1402)	$N(\pi \rightarrow \pi^*)$			4.49 (N)	4.52 (N)		4.41 (N_1)

[a] The excitation energies and oscillator strengths were calculated at the DFT/BP/TZP D_{4h} optimized geometry; [b] the excitation energies and oscillator strengths were calculated at the DFT/B3LYP/6–31G(d) D_{4h} optimized geometry; [c] data for ZnPc in gas-phase from [172]; [d] data for (CN)ZnPc, in DMA solution, at room temperature, from [174]; [e] data for (CN)ZnPc, in solution at cryogenic temperature, from [175], bold type is used for the Janh-Teller split components of the B_1 and B_2 bands; [f] data for ZnPc in Ar/matrix from [176]

almost entirely determined by the very large (4.17 au at SAOP level) transition dipole moment of the $2a_{1u}\rightarrow 7e_g$ transition. As for the higher excited states, it is apparent from the data in Table 19 that the excitation energies computed at B3LYP level are typically 0.5–0.6 eV higher than SAOP energies. The consequence is that the two approaches come to a somewhat different description of the B band region. A considerable experimental effort has been made to gain insight into the nature and number of excited states contributing to the B band. In the gas-phase spectrum reported by Edwards and Gouterman [172] the B band region is characterized by a broad intense and featureless absorption with the maximum located at 3.80 eV. A split of the B band in B_1 and B_2 components was clearly distinguished by Nyokong et al [174] in the absorption and MCD room temperature solution spectra of ZnPc(L) (L=cyanide, imidazole, pyridine). These authors invoked MCD spectra as an aid in the deconvolution analysis for the first time. In the region of interest they required the presence of two Faraday *A* terms for the fit of the MCD spectra, indicating that two degenerate excited states ("B_1 and B_2") exist in the B band region. In the case of zinc phthalocyanine with cyanide, they were located at 3.21 eV and 3.75 eV respectively. A second deconvolution analysis of the absorption and MCD spectra of ZnPc in Ar/matrix by VanCott et al [176] also suggests (at least) two absorptions, although with different characteristics. They propose one very broad absorption band with a maximum at 3.74 eV, which they call B_1. The onset of this band is identified as an observed origin at 3.05 eV that is followed by a progression of seven bands before the maximum at 3.74 eV. They believe that an experimental band maximum positioned at 3.71 eV does not fit into this pattern and corresponds to a separate electronic transition. On the basis of the MCD spectrum, they ascribe the strong broad band with maximum at 3.74 eV and the extra weaker band at 3.71 eV to two separate degenerate excited states and denote them as B_1 and B_2 respectively. More recently, Mack and Stillman [175] have reported a band deconvolution analysis of the absorption and MCD spectral data of (CN)ZnPc recorded at cryogenic temperature. They found that, in the B band region, the MCD spectrum could not be properly fitted using two *A* terms, as had been done in the earlier work by Nyokong et al [174] and VanCott et al [176], since a considerable residual intensity would not be accounted for. A satisfactory fit was obtained by replacing the *A* terms by two *B* terms of opposite sign with a slight separation (a pseudo-*A* term), which can be rationalized by a Jahn-Teller splitting of a degenerate state into two nondegenerate states which give rise to *B* terms. Mack and Stillman place the energy maxima of the two Janh-Teller split components of the excited state responsible for the B_1 band, which was previously assigned an energy of 3.21 eV, at 3.05 and 3.30 eV. Two more intense JT split components of the B_2 band, replacing the previous degenerate state at 3.75 eV, are placed at 3.57 and 3.81 eV in the deconvolution analysis. Between the two oppositely signed *B* terms comprising the pseudo-*A* fit of the B_1 band, an additional Faraday *B* term at 3.17 eV was introduced into the fit. As can be seen in Table 19, there is an excellent correspondence between the energies of the five bands of the deconvolution of Mack and Stillman and the five calculated TDDFT/SAOP ^1E_u transitions 3^1E_u–7^1E_u. The 6, 7^1E_u, which are the most intense ones, correspond to the experimental higher intensity in the B_2 region. The same holds for the three transitions in the B_1 region, 3, 4, 5^1E_u. In spite of the good agreement between calculated and experimental number and energy of the bands, and a reasonable agreement in intensities, there is a major discrepancy between the basis of TDDFT/SAOP results, namely unsplit degenerate states, and the underlying assumptions of the fit to experiment, namely the JT splitting of the degenerate states so that only nondegenerate excited states remain. In view of the generally very high quality of the TDDFT results for these types of systems, the authors consider it unlikely that the calculated number of states in this region (five ^1E_u and one $^1A_{2u}$) would not be correct. The calculated intensities do assign considerable intensity to the B_1 and B_2 regions and support this assumption. Calculation of JT splittings was beyond the scope of the investigation, so they were not able to make an assessment of the basic assumption of the fitting procedure, or to judge whether the fitting of the MCD spectrum with *A* terms for the degenerate states, as TDDFT/SAOP results would require, could be completely ruled out by the experimental data. It is clear however that, if JT splitting would have been significant, and experimentally visible in the splitting of the bands, it would have not been so large as to shift states out of the B region, so the fitting would then have to account for ten nondegenerate, pairwise very close-lying, states.

As inferred from Table 19, the description of the B_1/B_2 band region provided by TDDFT/SAOP is also in excellent agreement with the gas-phase spectrum and generally in line with deconvolution analyses of room temperature solution and Ar/matrix absorption and MCD spectra. With regard to the nature of the excited states accounting for the B_1/B_2 region, the 4^1E_u state which accounts for most of the intensity of the B_1 feature is mainly described by the $2b_{1u}\rightarrow 7e_g$ transition, while predominantly the $4a_{2u}\rightarrow 7e_g$ (Gouterman type) and $1a_{1u}\rightarrow 7e_g$ transitions enter in both the intense 6^1E_u and 7^1E_u corresponding to the experimentally higher intensity in the B_2 region.

In the B_1/B_2 region, B3LYP predicts only three ^1E_u excited states, the most intense of which (the 4^1E_u), computed at 3.74 eV with an oscillator strength of 0.328, is assigned to the B band main peak, which appears at 3.80 eV in the gas-phase spectrum, at 3.75 eV in the solution spectrum, and at 3.74 eV (B_1 max in Table 19) in the Ar/matrix spectrum. The 3^1E_u, computed at 3.66 eV, is assigned to the weaker B feature located at 3.21 eV (B_1) in the room temperature solution spectrum, and at 3.71 (B_2) in the Ar/matrix spectrum.

At variance with the SAOP results, B3LYP assigns the B band main peak to a state, the 4^1E_u, which is described by the $2b_{1u}{\rightarrow}7e_g$ and the weaker feature to the red side of the B band main peak to a state, the 3^1E_u, mainly composed of the $4a_{2u}{\rightarrow}7e_g$ non-Gouterman transition.

A relevant issue concerning the interpretation of the experimental spectra of ZnPc is the location of the n (N_b lone pairs)$\rightarrow\pi^*$ transitions. Analysis of Shpol'ski matrix data by Huang et al [178] and of Ar/matrix data by VanCott et al [176] suggested the presence of an additional n$\rightarrow\pi^*$ electronic transition at 2.08 eV, in the "Q_{02}" region of the spectrum of ZnPc. This band was called Q' and assigned to the e_u (N_b lone pairs)$\rightarrow e_g$ (LUMO) transition. Based on a deconvolution analysis of the absorption, and MCD spectral data of a vitrified solution of (CN)ZnPc, Mack and Stillman [179] have supported this additional *z*-polarized electronic band centered at 2.05 eV (604 nm) in the so called "Q_{02}" region, and also assigned this band to the e_u (N_b lone pairs)$\rightarrow e_g$ (LUMO) transition. Ar/matrix experiments by VanCott et al [176] locate a A_{2u} *z*-polarized n$\rightarrow\pi^*$ band labeled B_3 to the blue of the B band main peak (B_2), at 3.99 eV. In their solution spectra of ZnPc with cyanide and imidazole ligands, Nyokong et al [174] also observed a band in the vicinity of the B_3 absorption (at 4.13 eV for cyanide and 3.96 eV for imidazole), showing a pure Faraday *B* term. A weak band at 4.05 eV corresponding to a pure *B* term is also present in the deconvolution analysis of the B band performed by Mack and Stillman [175], although these authors could not distinguish whether this band was vibrational or electronic in origin.

As shown in Table 19, both B3LYP and SAOP calculations exclude the presence of a A_{2u} *z*-polarized n$\rightarrow\pi^*$ excited state in the tail of the Q band. The lowest A_{2u} state is indeed predicted at SAOP level at 3.28 eV and at B3LYP level at 3.99 eV. Actually, B3LYP assigns this state to the B_3 band. At variance with B3LYP, SAOP assigns the B_3 band to the 2^1A_{2u} excited state computed at 4.23 eV, which is a nearly pure $9b_{2g}$ (N_b lone pair)$\rightarrow$ $3b_{1u}$ (π^*) state.

As inferred from Table 20, where the TDDFT/SAOP excitation energies and oscillator strengths calculated for NiPc [137] are gathered together with the experimental data, TDDFT also proves to be a reliable approach for describing the electronic spectra of metallophthalocyanines in cases which are complicated by the presence of "additional" MLCT and LMCT bands. The main $\pi\rightarrow\pi^*$ features which characterize the visible and near-UV region (the Q, B_1, and B_2 bands), and the "extra" bands appearing in the UV region attributed by Edwards and Gouterman [172] to MLCT and LMCT states, are nicely accounted for by TDDFT calculations.

The domain of applicability of TDDFT to metallotetrapyrroles is not limited to ground state absorption spectra; it extends to the emission and excited state spectra as well.

Phosphorescence and triplet-triplet absorption spectra have been calculated for ZnPc at both the SAOP [135] and B3LYP [142] levels and for Al-

Table 20 Calculated excitation energies (eV) and oscillator strengths (in parentheses) for the optically allowed ^1E_u excited states of NiPc compared to experimental data

State	TDDFT/SAOP [137][a]		Expt[b]
1^1E_u	1.97 (0.6520)	Q	1.90 (Q)
2^1E_u	2.99 (0.0244)		
3^1E_u	3.16 (<0.0001)		
4^1E_u	3.20 (0.3396)	B_1	~3.40[c] (B_1)
5^1E_u	3.32 (0.0934)		
6^1E_u	3.40 (0.0160)		
7^1E_u	3.61 (0.2254)		
8^1E_u	3.85 (0.4650)	B_2	3.79 (B_2)
9^1E_u	3.91 (0.0952)	MLCT ($d_\pi \rightarrow \pi^*$)	4.37 (EB_1)
10^1E_u	4.02 (0.2162)		
11^1E_u	4.22 (0.5744)		
17^1E_u	4.94 (0.2596)	LMCT ($n \rightarrow d_{\sigma^*}$)	5.27 (EB_2)
18^1E_u	5.01 (0.3020)	MLCT/$\pi \rightarrow \pi^*$	
24^1E_u	5.48 (1.5268)	LMCT ($n \rightarrow d_{\sigma^*}$)	5.69 (EB_3)
25^1E_u	5.51 (0.1918)		
26^1E_u	5.58 (0.6334)	MLCT ($d_\pi \rightarrow \pi^*$)	

[a] The excitation energies and oscillator strengths were calculated at the DFT/BP/TZP D_{4h} optimized geometry; [b] gas-phase spectra of NiPc from [172]; [c] solution spectrum of NiPc in 1,2,4-trichlorobenzene from [209]

ClPc, MgPc, ZnP, ZnTPP, and ZnTBP at the B3LYP [142, 144] level. The calculated S_0–T_1 and T_1–T_n excitation energies have proven to be in excellent agreement with the experimental data, and are often more accurate than the corresponding S_0–S_n energies.

6.3 Sandwiches

Tetrapyrrole sandwich complexes have attracted considerable interest because they are good structural and spectroscopic models of the bacteriochlorophyll "special pair" in the reaction center of photosynthesis [180]. The metal ion in a sandwich complex holds the macrocycles closer than their van der Waals distance, resulting in strong π-π interactions which mimic the electronic interactions occurring within the special pair thought to be responsible for some of its peculiar properties, such as the ease of porphyrin π-system oxidation and the low energy of the first singlet (π, π^*) excited state compared to corresponding monomeric chromophores. This latter property produces a bathochromic shift in the long-wavelength absorption band and makes the dimer an effective trap for the harvested photon energy.

As a matter of fact, a number of synthetic sandwich complexes, both homoleptic (in which the two tetrapyrroles are identical) and heteroleptic (in which the two tetrapyrroles are different) exist to date, whose optical characteristics resemble those of the lowest excited states of the "special pair".

The homoleptic sandwiches have characteristic optical properties that include (i) momomer-like ground state absorption features, the Q and B bands, the latter being slightly blue shifted compared to the monomer, (ii) a new Q″ absorption band at higher energy, and a Q′ absorption band at lower energy than the monomer Q bands, (iii) a broad weak Q′ fluorescence band substantially red shifted from the Q′ absorption maximum, (iv) a phosphorescence band significantly red shifted from those of typical monoporphyrin complexes [181]. As for the optical spectra of heteroleptic sandwiches, these show, besides the Q, Q′, and Q″ features typical of the homoleptic complexes, two main features in the UV region that have been assigned to the Soret bands of the subunits on the basis of their correspondence to the Soret bands in homosandwiches [182, 183].

The essential features of the electronic spectra of homoleptic sandwiches have been interpreted on the basis of a simple supermolecule MO/CI model proposed by Holten et al [181, 184].

According to this model, the supermolecule MOs are formed by taking bonding and antibonding linear combinations of the monomer a_{1u} (π) and a_{2u} (π) HOMOs and e_g (π*) LUMOs of Gouterman's four-orbital model [154, 155, 185] that has been so successful in describing the electronic states of monoporphyrins. One-electron promotion among the eight orbitals of the sandwich results in four E_1 dipole-allowed and four E_3 dipole-forbidden excited state configurations in the pertinent D_{4d} symmetry. In the limit of degeneracy of the dipole-allowed configurations, (a_1e_1), (a_2e_1), (b_1e_3), and (b_2e_3), diagonalization of the CI matrix gives the singlet eigenfunctions B^+, Q^+, $CR_1{}^+$ and $CR_2{}^+$. The eigenfunctions B^+ and Q^+ are the allowed exciton states formed by in-phase combinations of the locally-excited (intra-ligand) B and Q states of monomer subunits, and $CR_1{}^+$ and $CR_2{}^+$ are the allowed charge resonance (CR) states formed by in-phase combinations of the inter-ligand charge transfer configurations. The pure exciton B^+ and Q^+ states account for the nearly unperturbed (with respect to the monoporphyrin) B and Q bands, whereas the CR^+ states account for the new Q″ band. The CR character of the Q″ band is consistent with the observed sensitivity of the energy and intensity of this band to the ionic radius of the metal ion (in other words, to the macrocycle separation). The four dipole-forbidden E_3 states have essentially mixed exciton and CR character. The lowest in energy, which is largely derived from the (a_1e_3) and (a_2e_3) configurations, is associated with the low-energy Q′ absorption.

The first accurate description of the optical properties of homoleptic tetrapyrrole sandwiches relies on a TDDFT study of zirconium(IV), cerium(IV) and thorium(IV) bis(porphyrins) [136].

This study leads to a reassessment of the accepted classification of the excited states and hence of the bands in the absorption spectrum, as purely excitonic excitations (intra-ring excitations that combine weakly with the equivalent one on the other ring) or purely charge-resonance excitations (ring-to-ring). As revealed by the ground-state electronic structure analysis, the highest occupied one-electron levels are composed of the porphyrin a_{1u} and a_{2u} orbitals (the Gouterman orbitals), but the level pattern is not simply a pair of low-lying nearly degenerate in-phase combinations and a pair of high-lying approximately degenerate antibonding combinations, as assumed in the eight orbital model. Instead, the a_{1u} orbitals on the two rings strongly overlap at the staggering angle of 45° and exhibit considerable splitting, while the a_{2u} orbitals overlap little and split little. Therefore, although the experimental spectra have porphyrin-like features, such as the well known Q and B bands, the actual composition of the states is rather different from that in porphyrins. In particular, the strong mixing of the $a_{1u}\rightarrow e_g^*$ and $a_{2u}\rightarrow e_g^*$ is absent, there is mixing with excitations of non-Gouterman type, and, in Ce, ring to metal charge transfer (RMCT) transitions play an important role.

Considering the theoretical interpretation of the optical spectra of the bis(porphyrin)M(IV) (M=Zr, Ce, Th) series given by Ricciardi et al [136] in more detail, it is apparent from Table 21 that TDDFT calculations achieve a very gratifying agreement with the experimental excitation energies. The very weak Q′ band of $Zr(OEP)_2$ and $Th(OEP)_2$ is assigned to the dipole forbidden 1^1E_3 state computed for the corresponding model systems, $Zr(P)_2$ and $Th(P)_2$, at 1.63 eV, and 1.78 eV respectively, in excellent agreement with the experiment. The assignment of the Q′ band to the 1^1E_3 state accounts very well for the observed breadth of this band and for the red shift (over 3000 cm^{-1}) relative to the $^1Q'$ absorption maximum of the broad weak $^1Q'$ fluorescence observed in zirconium, hafnium and thorium bis(porphyrin) sandwiches. According to TDDFT calculations, the 1^1E_3 state is largely derived from the promotion of one electron from the antibonding $5a_2$, (HOMO) to the bonding $15e_3$ (LUMO), so the potential energy curve of this state along the intermacrocycle distance coordinate is expected to be displaced to shorter bond distance, relative to the ground state, leading to a shift of the Franck-Condon intensity from the origin to higher vibronic components and a broad absorption contour.

The shift to longer wavelength of the 1^1E_3 state on going from $Th(P)_2$ to $Zr(P)_2$, as the splitting of the a_{1u}-derived $5a_2$ increases due to the shortening of the intermacrocycle distance, is also in line with the observed red shift of the Q′ band on going from the actinide to the transition metal bis(porphyrin) sandwiches. In the case of the cerium bis(porphyrin), because of the presence of low-lying 4*f* orbitals, the Q′ band has a considerable RMCT character. The transition from the antibonding $5a_2$ to the bonding e_g^* derived MO ($16e_3$) (which almost entirely accounts for the 1^1E_3 Q′ state in $Zr(P)_2$ and $Th(P)_2$), strongly mixes with the close lying $14e_3\rightarrow 15e_2$ RMCT transition in $Ce(P)_2$. Two nearly degenerate excited states result: the dipole forbid-

Table 21 Calculated excitation energies and oscillator strengths (f) for the optically allowed $^1E_1$ and $^1B_2$ and for the dipole forbidden $^1E_3$ excited states of $Zr(P)_2$, $Zr(OEP)_2$ (in parentheses), $Ce(P)_2$, and $Th(P)_2$ that contribute to the Q, Q′, Q″, and B bands, compared to the experimental data

TDDFT/BP [136]				Expt
State	Excitation energy (eV)	f	Assignment	
$Zr(P)_2$[a]				
1^1E_3	1.63 (1.44)		Q′	1.65[c] (Q′)
2^1E_1	2.10 (2.03)	0.0236 (0.0136)	Q	2.09[c] (Q)
3^1E_1	2.39 (2.33)	0.0001 (0.0029)	Q″	2.53[c] (Q″)
4^1E_1	2.85 (2.56)	0.0058 (0.0036)		
7^1E_1	3.39 (3.28)	1.5338 (1.9214)	B	3.24[c] (B)
$Ce(P)_2$[b]				
7^1E_3	1.78		Q′	~1.87[d]; ~1.93[e] (Q′)
8^1E_3	1.79			
4^1B_2	1.83	0.0302		
12^1E_1	2.13	0.0112	Q	2.16[d]; 2.16[e] (Q)
13^1E_1	2.34	0.0008	Q″	2.65[d]; 2.65[e] (Q″)
19^1E_1	2.85	0.0059		
26^1E_1	3.50	0.5418	B	3.28[d]; 3.28[e] (B)
$Th(P)_2$[b]				
1^1E_3	1.78		Q′	~1.74[f] (Q′)
2^1E_1	2.11	0.0083	Q	2.15[f]; 2.14[g] (Q)
3^1E_1	2.31	0.0002	Q_1''	~2.6[h] (Q_1''); 2.63 (sh)[g] (Q″)
4^1E_1	2.83	0.0052	Q_2''	~2.7[h] (Q_2'')
7^1E_1	3.52	2.0020	B	3.24[f]; 3.23[g] (B)

[a] The excitation energies and oscillator strengths were calculated at the DFT/BP/TZP D_{4d} optimized geometry; [b] the excitation energies and oscillator strengths were calculated using a combined scalar relativistic (SR) ZORA and TDDFT/BP approach at the (SR) ZORA D_{4d} optimized geometries; [c] data for $Zr(OEP)_2$ taken from dichloromethane solution from [187]; [d] data for $Ce(OEP)_2$ taken from cyclohexane solution from [188]; [e] data for $Ce(OEP)_2$ taken from dichloromethane solution from [210]; [f] data for $Th(OEP)_2$ taken from toluene solution from [211]; [g] data for $Th(OEP)_2$ taken from 3-methylpentane solution from [189]; [h] data for $Th(OEP)_2$ taken from 3-methylpentane glass at 78 K from [181]

den 7^1E_3 and 8^1E_3, calculated at 1.78 eV and 1.79 eV respectively. Dipole allowed excited states are also computed in the Q′ band region , the most intense of which, the 4^1B_2, is listed in Table 21, all having a predominant RMCT character. Dipole forbidden $^1E_3$ and weak dipole allowed $^1B_2$ and $^1E_1$ RMCT excited states are also predicted in the 700–1800 nm (1.8–0.7 eV) region (not shown in Table 21) where the absorption spectrum of $Ce(OEP)_2$ shows a weak tail. The energy of the RMCT transitions predicted by TDDFT calculations vary significantly with the exchange-correlation functional. At SAOP level, the RMCT transitions are indeed calculated at higher energy than at BP level, as a consequence of the upshift of 4*f* orbitals on going from

BP to SAOP. For instance, at SAOP level only two singlet E_3 excited states are predicted in the Q′ band region. The lower, at 1.78 eV, is largely described by the $5a_2 \rightarrow 16e_3$ transition with only 10% contribution from the $5b_1 \rightarrow 15e_3$ RMCT transition; the higher, at 1.87 eV, is an almost pure $5b_1 \rightarrow 16e_1$ RMCT state.

The presence of a cascade of singlet RMCT excited states in the Q′ band region and in the near IR, predicted at both BP and SAOP levels, nicely accounts for the rapid radiationless decay of $Ce(OEP)_2$ upon excitation with a 582-nm flash [186], and strongly supports the suggestion by Holten et al [186] that the ultrafast (1.5 ps) process that follows excitation of $Ce(OEP)_2$ is excited-state deactivation to the electronic ground state via low-lying RMCT states. As inferred from Table 21, the energy of the Q band is very well reproduced by TDDFT calculations, which assign this band to a $^1E_1$ state (2^1E_1 in Zr and Th, 12^1E_1 in Ce).

The four E_1 configurations of the eight orbital model, (a_2e_1), (a_1e_1), (b_1e_3), and (b_2e_3), all enter in the Q state, although with different weights. The important consequence is that, at variance with the model of Holten et al [181], the Q state is predicted to have a predominant CR character, rather than an excitonic character.

The solution spectra of $Zr(OEP)_2$ [187], $Ce(OEP)_2$ [188], and $Th(OEP)_2$ [181, 189] show a broad and weaker band to the blue of the Q band, denoted as Q″. At least two weak features account for the broad Q″ absorption, which are clearly distinguishable in the low temperature solution spectrum of $Th(OEP)_2$ [181]. TDDFT calculations by Ricciardi et al. [136] predict two ligand centered $^1E_1$ excited states in the Q″ region (the 3^1E_1 and 4^1E_1 in $Zr(P)_2$ and $Th(P)_2$, and the 13^1E_1 and 19^1E_1 in $Ce(P)_2$; in the case of $Ce(P)_2$ also several RMCT $^1E_1$ and $^1B_2$ excited states) which nicely reproduce the two features underlying the broad Q″ absorption, although the calculated oscillator strengths are somewhat underestimated. The discrepancy between the theoretical and experimental intensities was traced to the fact that the measured intensities refer to the alkylsubstituted compounds while the theoretical oscillator strengths refer to the bare $M(P)_2$ systems. Indeed, calculations performed on the octamethyl substituted zirconium bis(porphyrin) sandwich, $Zr(OMP)_2$ [136] show that the alkyl substituents have the effect of increasing the oscillator strength of the lower Q″ state which is predicted in $Zr(OMP)_2$ to be more than tenfold larger than in $Zr(P)_2$ (0.0118 vs 0.0001).

TDDFT calculations point to the fact that the splitting of the Q″ absorption is not related, as suggested by Holten et al [181] to the lifting of degeneracy of the four $^1E_1$ configurations involving the eight frontier orbitals; rather it is due to the presence of two excited states in this region, the higher-lying of which is dominated by a transition out of a non-Gouterman derived orbital to the bonding e_g* derived MO. The lower one contains only two of the four E_1 configurations, the (a_1e_1) and (b_1e_3), and, at variance with the model by Holten et al [181], has mixed EX/CR character in all members of the series.

According to the computed excitation energies and oscillator strengths, the B(0,0) band of $Zr(OEP)_2$ at 3.24–2.25 eV, and of $Th(OEP)_2$ at 3.23–3.24 eV, is assigned to the 7^1E_1 excited state, computed for $Zr(P)_2$ at 3.39 eV with an oscillator strength of 1.5338 and for $Th(P)_2$ at 3.52 eV with an oscillator strength of 2.002 (see Table 21). The B(0,0) band of $Ce(OEP)_2$ at 3.28 eV is very well accounted for by the 26^1E_1 excited state computed for $Ce(P)_2$ at 3.50 eV with an oscillator strength of 0.5418 (see Table 21). TDDFT calculations predict a pronounced multitransition character for the B state. The four transitions involving the orbitals of the eight orbital model all contribute, although with different weights, to the composition of this state, where they mix with other transitions: $\pi \rightarrow \pi^*$ (in all $M(P)_2$ systems), $\pi \rightarrow d_\sigma$ (in $Zr(P)_2$), $\pi \rightarrow f_\delta$ (in $Ce(P)_2$).

Analysis of the (approximate) wave function of the B state of $Zr(P)_2$ and $Th(P)_2$ in terms of excitonic, charge resonance, and RMCT contributions reveals that this state has a mixed EX/CR/RMCT character (66/15/19 in Zr, 68/17/14 in Th), although the excitonic parentage is by far the largest. This is not the case in $Ce(P)_2$, where the RMCT character is about 50%.

For the $M(P)_2$ series, a satisfactory description of the phosphorescence spectra and the triplet-triplet absorption spectra was also obtained by TDDFT calculations [136].

Besides the linear and nonlinear optical properties of bis(porphyrin) complexes of zirconium, cerium, and thorium, the optical spectra of bis(phthalocyanine) complexes have been investigated by TDDFT methods. In a combined experimental and theoretical study by Ercolani et al [139], the electronic spectra of Ti(IV) and Nb(IV) "stapled" bis(phthalocyanine) complexes have been computed via a TDDFT approach, leading to a satisfactory interpretation of their characteristic visible optical spectra. Although these species exhibit the typical features of tetrapyrrole sandwich complexes (the relative twisting and doming of the two phthalocyanine rings), the presence of the two inter-ligand C–C σ bonds has the important electronic effect of breaking down the delocalization of the π ring system. Therefore, the electronic spectra of the "stapled" bis(phthalocyanines) lack the typical features of the homoleptic tetrapyrrole sandwiches and are characterized instead by a broad feature extending from the red to the near-UV, which is responsible for the yellowish color of solutions of these systems.

TDDFT methods have also been applied successfully to the description of the linear and nonlinear optical properties of heteroleptic sandwich complexes. The optical spectrum and the hyperpolarizability of Zr(OEP)(OEPz,) for which large first hyperpolarizabilities, β^{SHG} (SHG=second-harmonic generation) were measured in an electric field induced second-harmonic generation (EFISH) experiment [182], have been investigated by TDDFT methods [134]. The excitation energies and oscillator strengths calculated

Table 22 Calculated excitation energies and oscillator strengths (f) for the optically allowed ^1E excited states of Zr(P)(Pz) that contribute to the Q, Q′, Q″, B_1, and B_2 bands, compared to experimental data

TDDFT/BP [134][a]				Expt[b]
State	Excitation energy (eV)	f	Assignment	
1^1E	1.39	0.0056	Q′	1.34 (Q′)
3^1E	2.02	0.0110	Q_1	2.07 (Q_1)
4^1E	2.33	0.0318	Q_2	2.25 (Q_2)
5^1E	2.43	0.0026	Q_{sh}	2.38 (Q_{sh})
9^1E	2.73	0.1638	Q″	2.88(Q″)
10^1E	3.04	0.0377	B_1	3.28 (B_1)
11^1E	3.15	0.0208		
12^1E	3.17	0.0492		
13^1E	3.19	0.0001		
14^1E	3.25	0.0044		
15^1E	3.42	0.0244	B_2	3.69(B_2)
16^1E	3.57	0.9820		
17^1E	3.68	0.1084		

[a] The excitation energies and oscillator strengths were calculated at the DFT/BP/TZP C_{4v} optimized geometry; [b] data for Zr(OEP)(OEPz) taken from dichloromethane solution from [182]

for the model complex, Zr(P)(Pz) [134], are gathered in Table 22 together with the experimental data concerning the octaethylsubstituted parent sandwich, Zr(OEP)(OEPz) [182].

As inferred from the data in Table 22, the salient features of the UV-vis spectrum of Zr(OEP)(OEPz) – the Q, Q′ and Q″ features, and the two main features in the UV region, the B_1 and B_2 bands – are well reproduced by TDDFT calculations. The very weak Q′ band centered at 1.34 eV is assigned to the 1^1E state calculated at 1.39 eV with an oscillator strength of 0.0056. The two absorptions at 2.07 eV (Q_1) and 2.25 eV (Q_2), and the associated higher energy shoulder at 2.38 eV (Q_{sh}), which characterize the composite Q band that extends from 2.0–2.6 eV in the solution spectrum of Zr(OEP)(OEPz) [182], are well accounted for by the 3^1E, 4^1E, and 5^1E excited states respectively. On the basis of the computed excitation energy and oscillator strength, the 9^1E state was believed to be responsible for the sharp and relatively intense Q″ band peaking at 2.88 eV. In the energy regime of the two B bands centered at 3.28 and 3.69 eV, TDDFT calculations predict two sets of closely spaced excited states, spanning the 2.91–3.26 eV and the 3.39–3.68 eV energy ranges. The summed oscillator strengths of 0.1630 and 1.1187 computed for the B_1 and B_2 bands agree nicely with the experimental relative intensities. However, the individual oscillator strengths indicate that it is the 16^1E excited state calculated at 3.57 eV that gives most of the intensity to the B band centered at 3.69 eV, whereas the intensity of the lower energy B band is dispersed over a number of excited states. A detailed analysis of the excited states in terms of intra-ligand (excitonic) and inter-ligand (charge transfer) transitions revealed that, due to the different electronic structure of the P and Pz subunits, the origin and the character of the Q, Q′, Q″, and B bands of the mixed P/Pz zirconium sandwich are different to the bis(porphyrin) analog. The usefulness of TDDFT for response properties has also been demonstrated in the investigation of $\beta(-2\omega; \omega; \omega)$ at various values of ω. It enabled the identification of the large negative β^{SHG} measured for Zr(OEP)(OEPz) at ω=0.65 eV in the EFISH experiment, a special result caused by the proximity of a pole in β (calculated at 0.69 eV). According to the frequency dependent TDDFT calculations on the model system, Zr(P)(Pz), the off-resonance values of β^{SHG} are not large. At low frequencies, including the static value (0.45×10^{-30} esu) they are positive and uninterestingly small, as expected for a molecule that does not have the strong donor-acceptor structure of the push-pull porphyrinic systems (which are among the best performing nonlinear optical (NLO) materials [190]), and is only slightly asymmetric.

7 Concluding Remarks

Undoubtedly, the most general conclusion to be drawn from this overview of recent TDDFT calculations on transition-metal complexes is that this technique, compared to other available theoretical methods, provides state-of-the-art results for excitation energies. We stress that in order for this to be a valid statement, there are two points that should receive due attention: the applied functional, and the geometry of the system.

Considering the functional, it is well-known that in the TDDFT calculations the orbital energies and the shapes of the Kohn-Sham orbitals are important. These are determined by the Kohn-Sham SCF potential. We have seen many examples where standard functionals like BP, PW91, B3LYP, PBE0, were definitely inferior to functionals for the Kohn-Sham potential such as LB94, LRC and SAOP, that have been specifically designed to overcome certain deficiencies that are known to exist for Kohn-Sham potentials derived from the GGA and hybrid exchange-correlation energy functionals. Such improvements (“shape corrections”) are mostly related to improved behavior in the interior region of the molecule, which lead to proper placement of the orbital energies. Asymptotic behavior also plays a role, but since it is particularly important for Rydberg orbitals, while we have been dealing here with valence excitations (metal-centered, ligand-centered, MLCT, or LMCT), correct asymptotic behavior is not as crucial here as it is for small molecules. Among the shape-corrected potentials, SAOP seems to perform best. In contrast, we have not found consistent improvement when moving from pure GGA functionals to hybrid functionals like B3LYP or PBE0. It seems that B3LYP performs at best at a level comparable to, say, BP, while in individual cases (like ZnPc, CrF_6) rather large differences with both standard GGAs and experiment occur. A quality comparable to CASPT2 and EOM-CCSD, for instance, is only achieved with the shape-corrected SAOP Kohn-Sham potential. All of the theoretical methods (CASPT2, EOM-CCSD, SAC-CI, SAOP-TDDFT) may still show occasional errors of several tenths of an eV (however, note the caveat below concerning assessment of the theoretical methods that ignores vibrational effects and vibronic interaction). Two cases can be identified where further improvement of functionals is mandatory. First, there is a problem with tight *d* levels, as shown by the poor results that have been reported for $CuCl_4^{2-}$ [106]. Second, the problem of charge transfer excitations [132] can distort the results, as has been found in systems where ligand-to-ligand charge transfer occurs [191]. We note that the many MLCT and LMCT transitions in TM complexes do not usually offer a problem, since in these systems charge transfer from or to the metal does not actually occur; instead they exhibit a remarkable constancy in the *d* population [94].

The second point to be stressed is the dependence of the calculated vertical excitation energies on the geometry of the complexes, in particular the metal-ligand distances. This is quite understandable, since excitations are out of or into molecular orbitals that often have strong metal-ligand bonding or antibonding character. An implication is that a DFT energy functional that yields a significant error in the bond length may for that reason already give a poor excitation energy (when calculated at the optimized geometry). The behavior of the functionals for the geometry is different for TM complexes than for the sets of small organic molecules (like G2) that have been used frequently as a training set for the parameters. For instance, the B3LYP functional does not perform particularly well for geometries of TM complexes. When judging the quality of TDDFT calculations of excitation energies, it is important to separate the effect of possible errors in the optimized geometry from errors arising from the functional in the actual calculation of the excitation energy. Given the considerable changes in vertical excitation energy with changes in the bond lengths, it is clear that vibrational averaging should be taken into account. In fact, we have been comparing to experimental vertical excitation energies that are usually simply obtained as the maxima of the absorption bands. It is however well-known that full vibronic interaction treatment may lead to band maxima that do not coincide with vertical excitation energies. In addition, for spectra in solution there are considerable solvent shifts.

The results discussed in this review indicate that the agreement of TDDFT with experiment is not as good for TM complexes as has been observed for small benchmark molecules such as N_2, CO, CH_2O, and furan, where an accuracy of 0.1 – 0.2 eV has often been obtained. The best TDDFT methods for absorption spectra of TM complexes often appear to have errors of a few tenths of an eV. However, it is evident that a full assessment of the quality of TDDFT (as well as other theoretical methods) for TM complexes can only be given when geometry effects have been incorporated via a full vibronic interaction treatment.

References

1. Pierloot K (2003) Mol Phys 101(13):2083
2. Daniel C (2003) Coord Chem Rev 238:143
3. Ziegler T, Rauk A, Baerends EJ (1977) Theoret Chim Acta 43:261
4. Slater JC (1960) Quantum theory of atomic structure, vol 1. McGraw-Hill, New York
5. Eyring H, Walter J, Kimball GE (1944) Quantum chemistry. Wiley, New York
6. Runge E, Gross EKU (1984) Phys Rev Lett 52:997
7. Gross EKU, Kohn W (1990) Adv Quantum Chem 21:255
8. Gross EUK, Dobson JF, Petersilka M (1996) In: Nalewajski RF (ed) Density functional theory. Springer Series: Topics in Current Chemistry. Springer, Berlin Heidelberg New York
9. Casida ME (1995) In: Chong DP (ed) Recent advances in density functional methods, vol 1. World Scientific, Singapore, p 155
10. Bauernschmitt R, Ahlrichs R (1996) Chem Phys Lett 256: 454
11. van Gisbergen SJA, Snijders JG, Baerends EJ (1995) J Chem Phys 103: 9347
12. van Gisbergen SJA, Osinga VP, Gritsenko OV, van Leeuwen R, Snijders JG, Baerends EJ (1996) J Chem Phys 105: 3142
13. Osinga VP, van Gisbergen SJA, Snijders JG, Baerends EJ (1997) J Chem Phys 106: 5091
14. Hohm U, Goebel D, Grimme S (1997) Chem Phys Lett 272: 1059
15. Görling A, Heinze HH, Ruzankin SP, Staufer M, Rösch N (1999) J Chem Phys 110: 2785
16. Ioannou AG, Colwell SM, Amos RD (1997) Chem Phys Lett 278: 278
17. Grüning M, Gritsenko OV, van Gisbergen SJA, Baerends EJ (2002) J Chem Phys 116: 9591
18. Dreizler RM, Gross EKU (1990) Density functional theory: an approach to the many-body problem. Springer, Berlin Heidelberg New York
19. Jamorski C, Casida ME, Salahub DR (1996) J Chem Phys 104:5134
20. Petersilka M, Gossmann UJ, Gross EKU (1996) Phys Rev Lett 76:1212
21. Bauernschmitt R, Häser M, Treutler O, Ahlrichs R (1997) Chem Phys Lett 264:573
22. Tozer DJ, Handy NC (1998) J Chem Phys 109:10180
23. Stratmann RE, Scuseria GE, Frisch MJ (1998) J Chem Phys 109:8218
24. van Gisbergen SJA, Koostra F, Schipper PRT, Gritsenko OV, Snijders JG, Baerends EJ (1998) Phys Rev A 57:2556
25. Casida ME, Salahub DR (2000) J Chem Phys 113:8918
26. Schipper PRT, Gritsenko OV, van Gisbergen SJA, Baerends EJ (2000) J Chem Phys 112:1344
27. Zangwill A, Soven P (1980) Phys Rev 21:1561
28. Deb BM, Ghosh SK (1983) Int J Quantum Chem 23:1
29. Stott MJ, Zaremba E (1980) Phys Rev 21:12
30. Mahan GD (1980) Phys Rev A 22:1780

31. Bartolotti L, Acharya PK (1982) J Chem Phys 77:4576
32. van Gisbergen SJA (1998) Molecular response property calculations using time-dependent density functional theory. PhD Thesis, Vrije Universiteit, Amsterdam
33. Oddershede J, Jorgensen P, Yeager DL (1984) Comput Phys Rep 2:33
34. Petersilka M, Gross EKU (1996) Int J Quantum Chem Symp 30:181
35. Almbladh CO, Pedroza AC (1984) Phys Rev A 29:2322
36. Aryasetiawan F, Stott MJ (1986) Phys Rev B 34:4401
37. Zhao Q, Morrison RC, Parr RG (1994) Phys Rev A 50:2138
38. van Leeuwen R, Baerends EJ (1994) Phys Rev A 49:2421
39. Gritsenko OV, van Leeuwen R, Baerends EJ (1995) Phys Rev A 52:1870
40. Ingamells VE, Handy NC (1996) Chem Phys Lett 248:373
41. Schipper PRT, Gritsenko OV, Baerends EJ (1997) Theor Chem Acc 98:16
42. Schipper PRT, Gritsenko OV, Baerends EJ (1998) Theor Chem Acc 99:329
43. Vosko SH, Wilk L, Nusair M (1980) Can J Phys 58:1200
44. Becke A (1988) Phys Rev A 38:3098
45. Perdew JP (1986) Phys Rev B 33:8822 (Erratum: PRB 34 (1986) 7406)
46. Petersilka M, Gossmann UJ, Gross EKU (1998) Electronic density functional theory: recent progress and new directions. Plenum, New York
47. Talman JD, Shadwick WF (1976) Phys Rev A 14:36
48. Casida ME, Casida KC, Salahub DR (1998) Int J Quantum Chem 70:933
49. Grüning M, Gritsenko OV, van Gisbergen SJA, Baerends EJ (2001) J Chem Phys 114:652
50. Chong DP, Gritsenko OV, Baerends EJ (2002) J Chem Phys 116:1760
51. Gritsenko OV, Baerends EJ (2002) J Chem Phys 117:9154
52. Gritsenko OV, van Leeuwen R, van Lenthe E, Baerends EJ (1995) Phys Rev A 51:1944
53. Hamprecht FA, Cohen AJ, Tozer DJ, Handy NC (1998) J Chem Phys 109:6264
54. Stratman RE, Scuseria GE, Frisch MJ (1996) Chem Phys Lett 257:213
55. Becke AD (1993) J Chem Phys 98:5648
56. Lee C, Yang W, Parr RG (1988) Phys Rev B 37:785
57. Perdew JP, Burke K, Ernzerhof M (1996) Phys Rev Lett 77:3865
58. Perdew JP, Burke K, Ernzerhof M (1997) Phys Rev Lett 78:1396
59. Adamo C, Barone V (1999) J Chem Phys 110:6158
60. Beach NA, Gray HB (1968) J Am Chem Soc 90:5731
61. Lever ABP, Ozin GA, Hanlan AJL, Power WJ, Gray HB (1979) Inorg Chem 18:2088
62. Levenson RA, Gray HB, Ceasar GP (1970) J Am Chem Soc 92: 3653
63. Levenson RA, Gray HB (1975) J Am Chem Soc 97:6042
64. Koerting CF, Walzl KN, Kupperman A (1987) J Chem Phys 86:6646
65. Kotzian M, Rösch N, Schröder H, Zerner MC (1989) J Am Chem Soc 111:7687
66. Bauernschmitt R, Ahlrichs R, Hennrich FH, Kappes MM (1998) J Am Chem Soc 120: 5052
67. van Gisbergen SJA, Rosa A, Ricciardi G, Baerends EJ (1999) J Chem Phys 111:2505
68. van Gisbergen SJA, Groeneveld JA, Rosa A, Snijders JG, Baerends EJ (1999) J Phys Chem A 103:6835
69. Rosa A, Baerends EJ, van Gisbergen SJA, van Lenthe E, Groeneveld JA, Snijders JG (1999) J Am Chem Soc 121:10356
70. Pierloot K, Tsokos E, Vanquickenborne LG (1996) J Phys Chem 100:16545
71. Kühn O, Hachey MRD, Rohmer MM, Daniel C (2000) Chem Phys Lett 322:199
72. Hada M, Imay Y, Hidaka M, Nakatsuji H (1995) J Chem Phys 103:6993
73. Rosa A, Ricciardi G, Baerends EJ, Stufkens DJ (1995) Inorg Chem 34:3425
74. Pollak C, Rosa A, Baerends EJ (1997) J Am Chem Soc 119:7324

75. Baerends EJ, Rosa A (1998) Coord Chem Rev 177:97
76. Schreiner AF, Brown TL (1968) J Am Chem Soc 90:3366 (solution spectrum of $Ni(CO)_4$)
77. Gray HB, Beach N (1963) J Am Chem Soc 85:2922
78. Pierloot K, Verhulst J, Verbeke P, Vanquickenborne LG (1989) Inorg Chem 28:3059
79. Geoffroy GL, Wrighton MS (1979) Organometallic photochemistry. Academic, New York
80. Ferraudi GJ (1988) Elements of inorganic photochemistry. Wiley, New York
81. Rosa A, Ehlers AW, Baerends EJ, Snijders JG, te Velde G (1996) J Phys Chem 100:5690
82. Martin M, Rees B, Mitschler A (1982) Acta Cryst B 38:6
83. Almenningen A, Jacobsen GG, Seip HM (1969) Acta Chem Scand 23:685
84. Boulet P, Chermette H, Daul C, Gilardoni F, Rogemond F, Weber J, Zuber G (2001) J Phys Chem A 105:885
85. Lembarki A, Rogemond F, Chermette H (1995) Phys Rev A 52:3704
86. Swart M, Snijders JG (2003) Theor Chem Acc 110:34
87. Daul C, Güdel H-U, Weber J (1993) J Chem Phys 98:4023
88. Full J, González L, Daniel C (2001) J Phys Chem A 105:184
89. Nakai H, Ohmori Y, Nakatsuji H (1991) J Chem Phys 95:8287
90. Buijse MA, Baerends EJ (1990) J Chem Phys 93:4129
91. Johnson KH, Smith JFC (1971) Chem Phys Lett 10:219
92. Hillier IH, Saunders VR (1971) Chem Phys Lett 9:219
93. Johansen H, Rettrup S (1983) Molec Phys 49:1209
94. Ziegler T, Rauk A, Baerends EJ (1976) Chem Phys 16:209
95. Dickson RM, Ziegler T (1996) Int J Quantum Chem 58:681
96. Stückl AC, Daul C, Güdel HU (1997) Int J Quantum Chem 61:579
97. Nooijen M, Lotrich V (2000) J Chem Phys 113:494
98. Holt SL, Ballhausen CJ (1967) Theor Chim Acta 7:313
99. Nemykin VN, Basu P (2003) Inorg Chem 42:4046
100. Hellberg KH, Müller A, Glemser O (1963) Z Naturforsch B 21:188
101. Hope EG, Jones PJ, Levason W, Ogden JS, Tajok M, Turff JW (1985) J Chem Soc Dalton 1443
102. Adamo C, Barone V (2000) Theor Chem Acc 105:169
103. Nagle J, Baerends EJ (unpublished results)
104. Nagle J (2002) private communication
105. Nakai H, Morita H, Tomasello P, Nakatsuji H (1998) J Phys Chem A 102:2033
106. Szilagyi RK, Metz M, Solomon EI (2002) J Phys Chem A 106:2994
107. Autschbach J, Jorge FE, Ziegler T (2003) Inorg Chem 42:2867
108. Klamt A, Jonas V (1996) J Chem Phys 105:9972
109. Pye CC, Ziegler T (1999) Theor Chem Acc 101:396
110. Ziegler T (2002) J Chem Soc Dalton 642
111. Furche F, Ahlrichs R, Wachsmann C, Weber E, Sobanski A, Vögtle F, Grimme S (2000) J Am Chem Soc 122:1717
112. Cheeseman JR, Frisch M, Devlin FJ, Stephens PJ (2000) J Phys Chem A 104:1039
113. Autschbach J, Ziegler T, van Gisbergen SJA, Baerends EJ (2002) J Chem Phys 116:6930
114. Hagfeldt A, Graetzel M (2000) Acc Chem Res 33:269
115. Rudmann H, Shimada S, Rubner MF (2002) J Am Chem Soc 124:4918
116. Pierloot K, Ceulemans A, Merchan M, Serrano-Andres L (2000) J Phys Chem A 104:4374

117. Halls MD, Schlegel HB (2001) Chem Mater 13:2632
118. Zgiersky MZ (2003) J Chem Phys 118:4045
119. Chen LX, Shaw GB, Novozhilova I, Liu T, Jennings G, Attenkofer K, Meyer GJ, Coppens P (2003) J Am Chem Soc 125:7022
120. Guillemoles J-F, Barone V, Joubert L, Adamo C (2002) J Phys Chem A 106:11354
121. Fantacci S, De Angelis F, Selloni A (2003) J Am Chem Soc 125:4381
122. Gorelsky SI, Lever ABP (2001) J Organomet Chem 635:187
123. Hay PJ (2002) J Phys Chem A 106:1634
124. Klein A, van Slageren J, Zalis S (2003) Eur J Inorg Chem 1917
125. Shklover V, Nazeeruddin MK, Zakeeruddin SM, Barbè C, Kay A, Haibach T, Steurer W, Hermann R, Nissen H-U, Grätzel M (1997) Chem Mater 9:430
126. Nazeeruddin MK, Zakeeruddin SM, Humpry-Baker R, Jirousek M, Liska P, Vlachopoulos N, Shklover V, Fisher C-H, Grätzel M (1999) Inorg Chem 38:6298
127. Reichardt C (1994) Chem Rev 94:2319
128. Tomasi J, Persico M (1994) Chem Rev 94:2027
129. Klamt A, Schürmann G (1993) J Chem Soc Perk T 2 799
130. Aarnts MP, Wilms MP, Peelen K, Fraanje J, Goubitz K, Hartl F, Stufkens DJ, Baerends EJ, Vlcek A Jr (1996) Inorg Chem 35:5468
131. Aarnts MP, Stufkens DJ, Wilms MP, Baerends EJ, Vlcek A Jr, Klark IP, George MW, Turner JJ (1996) Chem Eur J 2:1556
132. Dreuw A, Weisman JL, Head-Gordon M (2003) J Chem Phys 119:2943
133. Sundholm D (1999) Chem Phys Lett 302:480
134. Ricciardi G, Rosa A, van Gisbergen SJA, Baerends EJ (2000) J Phys Chem A 104:635
135. Ricciardi G, Rosa A, Baerends EJ (2001) J Phys Chem A 105:5242
136. Ricciardi G, Rosa A, Baerends EJ, van Gisbergen SJA (2002) J Am Chem Soc 124:12319
137. Rosa A, Ricciardi G, Baerends EJ, van Gisbergen SJA (2001) J Phys Chem A 105:3311
138. Baerends EJ, Ricciardi G, Rosa A, van Gisbergen SJA (2002) Coord Chem Rev 230:5
139. Bauer EM, Donzello MP, Ercolani C, Masetti E, Panero S, Ricciardi G, Rosa A, Chiesi-Villa A, Rizzoli C (2002) Inorg Chem 42:283
140. Nguyen KA, Pachter R (2001) J Chem Phys 114:10757
141. Nguyen KA, Day PN, Pachter R (1999) J Chem Phys 110:9135
142. Nguyen KA, Kennel J, Pachter R (2002) J Chem Phys 117:7128
143. Wasbotten IH, Wondimagegn T, Ghosh A (2002) J Am Chem Soc 124:8104
144. Nguyen KA, Pachter R (2003) J Chem Phys 118:5802
145. Kyrychenko A, Andréasson J, Mårtensson J, Albinsson B (2003) J Phys Chem B
146. Dreuw A, Dunietz BD, Head-Gordon M (2002) J Am Chem Soc 124:12070
147. Dunietz BD, Dreuw A, Head-Gordon M (2003) J Phys Chem B 107:5623
148. Liu X, Xu L-C, He T-J, Chen D-M, Liu F-C (2003) Chem Phys Lett 379:517
149. Hasegawa J, Hada M, Nonoguchi M, Nakatsuji H (1996) Chem Phys Lett 250:159
150. Rubio M, Roos BO, Serrano-Andrés L, Merchán M (1999) J Chem Phys 15:7202
151. Hashimoto T, Choe Y-K, Nakano H, Hirao K (1999) J Phys Chem A 103:1894
152. Kadish KM, Smith KM, Guilard R (2000) The porphyrin handbook. Academic, New York
153. Edwards L, Dolphin DH, Gouterman M (1970) J Mol Spectrosc 35:90
154. Gouterman M (1959) J Chem Phys 30:1139
155. Gouterman M (1961) J Mol Spectrosc 6:138
156. Gouterman M, Wagnière G, Snyder LC (1963) J Mol Spectrosc 11:108
157. Simpson WT (1949) J Chem Phys 17:1218

158. Platt JR (1956) In: Hollaender A (ed) Radiation biology, vol 3. McGraw Hill, New York, p 71
159. Sekino H, Kobayashi H (1987) J Chem Phys 86:5045
160. Pileni MP, Gratzel M (1980) J Phys Chem 84:1822
161. Keegan JD, Stolzenberg AM, Lu Y-C, Linder RE, Barth G, Moscowitz A, Bunnenberg E, Djerassi C (1982) J Am Chem Soc 104:4305
162. Canters GW, Jansen G, Noort M, van der Waals JH (1976) J Phys Chem 80:2253
163. Makinen MW, Eaton W (1973) Ann NY Acad Sci 206:210
164. Tokita Y, Nakatsuji H (1997) J Phys Chem B 101:3281
165. Weiss C, Kobayashi H, Gouterman M (1965) J Mol Spectrosc 16:415
166. McHugh AJ, Gouterman M, Weiss C (1972) Theor Chim Acta 24:346
167. Wagner RW, Lindsey JS, Seth J, Palaniappan V, Bocian DF (1996) J Am Chem Soc 118:3996
168. Reimers JR, Lü TX, Crossley MJ, Hush NS (1996) Chem Phys Lett 256:353
169. Leznoff CC, Lever ABP (1990-1996) In: Leznoff CC, Lever ABP (eds) Phthalocyanines: properties and applications, vols 1–4. VCH, New York
170. Fernández DA, Awruch J, Dicelio LE (1996) Photochem Photobiol 63:784
171. Lawrence DS, Whitten DG (1996) Photochem Photobiol 64:923
172. Edwards L, Gouterman M (1970) J Mol Spectrosc 33:292
173. Schechtman BH, Spicer EW (1970) J Mol Spectrosc 33:28
174. Nyokong TN, Gasyna Z, Stillman MJ (1987) Inorg Chem 26:1087
175. Mack J, Stillman MJ (1995) J Phys Chem 99:7935
176. VanCott TC, Rose JL, Misener GC, Williamson BE, Schrimpf AE, Boyle ME, Schatz PN (1989) J Phys Chem 93:2999
177. Metcalf DH, VanCott TC, Snyder SW, Schatz PN, Williamson BE (1990) J Phys Chem 94:2828
178. Huang T-H, Rieckhoff KE, Voigt EM (1982) J Chem Phys 77:3424
179. Mack J, Stillman MJ (1997) Inorg Chem 36:413
180. Deisenhofer J, Michel H (1989) Science 245:1463
181. Bilsel O, Rodriguez J, Milam SN, Gorlin PA, Girolami GS, Suslick KS, Holten D (1992) J Am Chem Soc 114:6528
182. Collman JP, Kendall JL, Chen JL, Eberspacher TA, Moylan CR (1997) Inorg Chem 36:5603
183. Guilard R, Barbe J-M, Ibnlfassi A, Zrineh A, Adamian VA, Kadish KM (1995) Inorg Chem 34:1472
184. Bilsel O, Buchler JW, Hammerschmitt P, Rodriguez J, Holten D (1991) Chem Phys Lett 182:415
185. Gouterman M (1978) In: Dolphin D (ed) The porphyrins, Vol III, Part A. Academic, New York, p 1
186. Bilsel O, Rodriguez J, Holten D (1990) J Phys Chem 94:3508
187. Buchler JW, De Cian A, Elschner S, Fischer J, Hammerschmitt P, Weiss R (1992) Chem Ber 125:107
188. Buchler JW, De Cian A, Fischer J, Kihn-Botulinski M, Paulus H, Weiss R (1986) J Am Chem Soc 108:3652
189. Kadish KM, Moninot G, Hu Y, Dubois D, Ibnlfassi A, Barbe J-M, Guilard R (1993) J Am Chem Soc 115:8153
190. Le Cours SM, Guan H-W, Di Magno SG, Wangand CH, Therien MJ (1996) J Am Chem Soc 118:1497
191. Turki M, Daniel C, Zális S, Vlcek A Jr, van Slageren J, Stufkens DJ (2001) J Am Chem Soc 123:11431

192. Chang C, Pelissier M, Durand P (1986) Phys Scripta 34:394
193. Heully J-L, Lindgren I, Lindroth E, Lundqvist S, Màrtensson-Pendrill A-M (1986) J Phys B 19:2799
194. van Lenthe E, Baerends EJ, Snijders JG (1993) J Chem Phys 99:4597
195. van Lenthe E, Baerends EJ, Snijders JG (1994) J Chem Phys 101:9783
196. van Lenthe E, Ehlers AW, Baerends EJ (1999) J Chem Phys 110:8543
197. Sohn YS, Hendrickson DN, Gray HB (1971) J Am Chem Soc 93:3603
198. Giordano PJ, Wrighton MS (1977) Inorg Chem 16:160
199. Palenik GJ (1967) Inorg Chem 6:503
200. Johnson IW, McGlynn SP (1970) Chem Phys Lett 7:618
201. Muller A, Diemann E, Jorgensen CK (1973) Struct Bond 14:23
202. Ortner MH (1961) J Chem Phys 34:556
203. McDiarmid R (1974) J Chem Phys 61:3333
204. Seip HM, Seip R (1966) Acta Chem Scand 20:2698
205. Rosa A, Ricciardi G, Baerends EJ, van Gisbergen SJA (1999) In: Proc 8th Int Conf Appl Density Functional Theory to Chemistry and Physics, Rome, Italy, p 38
206. Starukhin A, Shulga A, Waluk J (1997) Chem Phys Lett 272:405
207. Jentzen W, Turowska-Tyrk I, Scheidt WR, Shelnutt JA (1996) Inorg Chem 35:3559
208. Dvornikov SS, Knyukshto VN, Kuzmitskii VA, Shulga AM, Solovyov KN (1981) J Lumin 23:373
209. Bonamico M, Fares V, Flamini A, Poli N (1991) Inorg Chem 30:3081
210. Donohoe RJ, Duchowski JK, Bocian DF (1988) J Am Chem Soc 110:6119
211. Bilsel O, Rodriguez J, Holten D, Girolami GS, Milam SN, Suslick KS (1990) J Am Chem Soc 112:4075

Structure and Bonding (2004) 112:117–149
DOI 10.1007/b97938

Applications of Hybrid DFT/Molecular Mechanics to Homogeneous Catalysis

Gregori Ujaque · Feliu Maseras (✉)

Unitat de Química Física, Edifici Cn., Universitat Autónoma de Barcelona, 08193 Bellaterra, Catalonia, Spain
gregori@klingon.uab.es; feliu@klingon.uab.es

Abstract The applications of hybrid DFT/molecular mechanics (DFT/MM) methods to the study of reactions catalyzed by transition metal complexes are reviewed. Special attention is given to the processes that have been studied in more detail, such as olefin polymerization, rhodium hydrogenation of alkenes, osmium dihydroxylation of alkenes and hydroformylation by rhodium catalysts. DFT/MM methods are shown, by comparison with experiment and with full quantum mechanics calculations, to allow a reasonably accurate computational study of experimentally relevant problems which otherwise would be out of reach for theoretical chemistry.

Keywords Homogeneous Catalysis · Quantum Mechanics/Molecular Mechanics · Polymerization · Hydrogenation · Dihydroxylation

1 Introduction

Homogeneous catalysis is a very important research field within inorganic chemistry [1, 2]. Transition metal complexes are able to catalyze a large number of organic reactions, and they therefore provide the key to the synthesis of many industrially relevant products. Homogeneous catalysis is not yet economically as important in terms of sheer size as its heterogeneous alternative, but its relative use has been growing steadily for the last 40 years.

The preparation of catalysts is often easier in heterogeneous catalysis, and products are definitely easier to separate. Nevertheless, homogeneous catalysis has the substantial advantage of typically requiring milder conditions and yielding a higher selectivity. This increases the purity of the products and reduces the amount of waste, two concepts which are very useful considering the desires to develop a more environmentally friendly chemical industry.

The transition metal complexes used in practice as catalysts are often quite large, easily in the range of hundreds of atoms. This poses a challenge to computational chemistry. Accurate DFT methods [3] including the generalized gradient approximation (GGA) are often required for the description of the transition metal center and its immediate environment, but they are too time consuming for the large experimental systems. Other methods capable of dealing with large systems, like standard force fields in molecular mechanics (MM) [4], are unable to properly handle the complexities associated with the metal center. A possible solution to this problem is the performance of accurate calculations on model systems. In this way, PH_3 represents any phosphine, and C_5H_5 represents any cyclopentadienyl ring. This approach has been very successful for determining the general mechanisms of a lot of catalytic processes [5–7], but it is obviously unable to differentiate between the behaviors of systems which differ in some substituents far away from the metal center, and this is often critical for selectivity. A more sophisticated solution is the partition of the system into two regions, and the use of a different description for each region. In this way, the computationally demanding quantum mechanical (QM) description is only applied in the region where it is strictly necessary. This is precisely what QM/MM methods do. DFT/MM is just a particular case of QM/MM.

QM/MM methods have been around for some time in computational chemistry, with the first proposal of this approach having been made in the 1970s [8]. Initial applications throughout the 1980s and the early 1990s were mostly concerned with the introduction of solvent effects, with special focus on biochemical problems. The solvent is often in the MM region, and its electronic effect on the QM region is usually described by point charges. The handling of covalent bonds across the QM/MM partition within this approach is not trivial, but these methods have reached a state of maturity which makes them quite widespread nowadays in computational biochemistry [9, 10].

The situation for transition metal chemistry has been somehow different because, while the description of the electronic effects from the MM region may not be that important, the handling of covalent connections between the QM and MM regions is critical. The methods that have been most successful in handling this situation have been the closely related IMOMM [11] and ONIOM [12, 13]. These schemes provide a computationally economical and methodologically robust method of introducing the steric effects of the

MM region into the calculation, although they are less precise about the electronic effects, which are directly neglected in the most simple version. The application of QM/MM methods is not limited to biochemical systems and transition metal chemistry; a number of works have also been carried out on heterogeneous catalysis [14].

Despite their relatively recent appearance in 1995, the IMOMM and ONIOM methods have been extremely productive when applied to transition metal chemistry [15]. A substantial amount of publications have been concerned with homogeneous catalysis, and they will be reviewed in this contribution. Other applications to transition metal chemistry, in structure [16, 17] and reactivity [18, 19], will not be treated here. Methodological aspects of the QM/MM approach, in connection with its application to transition metal chemistry, have been reviewed in a number of papers [20–22].

The present contribution is organized by catalytic processes. Each reaction that has been the subject of at least three published DFT/MM works has been allotted a separate section. Other works are collected together in a single section. A final section presents the concluding remarks.

2 Olefin Polymerization

The catalytic polymerization of olefins is a reaction of vast economic importance. Polyolefins have a large and increasing application potential because they contain only carbon and hydrogen atoms, are inert, stable to water, and can be easily recycled or used as a source of energy through incineration [23]. The reaction is widely used on an industrial scale with heterogeneous Ziegler-Natta catalysts, which are usually early transition metal halides activated by cocatalysts such as alkylaluminum compounds, following the original proposals of the groups of Ziegler and Natta in the 1950s [24, 25].

Homogeneous variations of this process are also industrially relevant. The practical application of the homogeneous process started with the work by Kaminsky and co-workers in the 1970s [26, 27]. These authors discovered the high catalytic activity of mixtures of early transition metal complexes containing metallocene ligands and methylalumoxane (MAO) as cocatalyst. Homogeneous catalysis allows a much better control of the microscopic structure of the polymers by tuning of the ligands. Research in this field continues to be very active [28]. The range of catalysts able to carry olefin polymerization has been further expanded more recently to other complexes containing different set of ligands and metal centers [29, 30].

The reaction mechanism of catalyzed olefin polymerization looks deceptively simple [31], and is shown in Fig. 1. The catalyst originally has an alkyl ligand and a vacant site. The olefin first coordinates by η^2 to the vacant site,

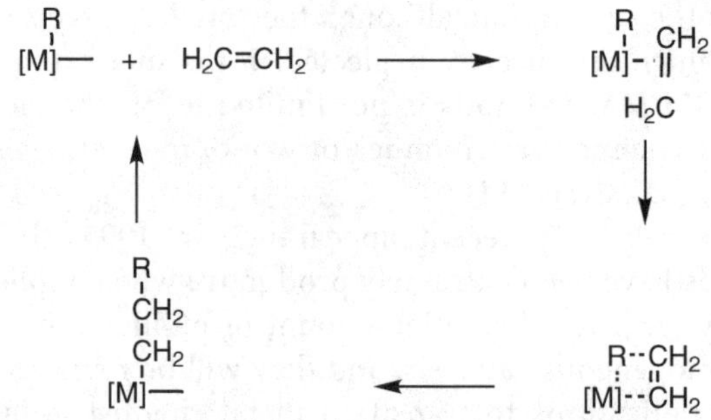

Fig. 1 Generally accepted mechanism for the polymerization of ethylene

and it later inserts into the metal-alkyl bond, yielding a complex with a new (and longer) alkyl group and a vacant site. This resulting complex is ready to start the next catalytic cycle with a new olefin molecule. A more detailed analysis of the mechanism hints however at its real complexity. In first place, the role of the cocatalysts (such as MAO) in the activation of the catalyst is not trivial. The length of the polymer chain, critical for its properties, will depend on the relative energy barrier of the termination reaction, and there are different possibilities for this. The propagation reaction depicted in Fig. 1 would always yield the linear product, but branched polymers are also accessible with the introduction of additional β C-H oxidation/elimination reactions. Finally, if the olefin has some substituent, possibilities for regio- and stereoselectivity appear.

Remarkably enough, a great deal of these mechanistical complexities have been successfully used for the tailoring of catalysts suited for the production of polymers with particular properties. Detailed knowledge has been gained throughout the years about the reaction mechanisms of a variety of catalysts, and theoretical calculations have contributed substantially to this knowledge [32–34]. Early works were carried out on simple models [35], but the mechanistic subtleties of the process can only be fully appreciated when the full catalyst is introduced into the calculation. Its large size makes this topic highly appropriate for DFT/MM methods. In fact, as we are writing this review, this is easily the chemical process that has been treated most often with these methods. Note that our overview of calculations will be organized by mechanistical aspect rather than by compound.

We will start by discussing the calculations on ethylene polymerization. In this case, the issues are mostly polymer length and the formation of linear or branched polymers. The first DFT/MM applications on this topic were published near simultaneously in 1997 and 1998 by the groups of Ziegler [36] and Morokuma [37, 38] on the reaction of ethylene polymerization catalyzed by cationic diimine Ni(II) complexes of the type (ArN=C(R)-C(R)=NAr)-Ni-CH_3^+. Shortly before this, these species (one of them is

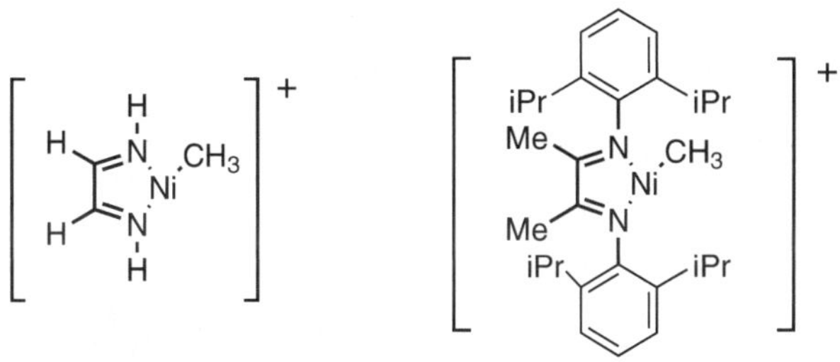

Fig. 2 DFT model (left) and full DFT/MM (right) descriptions commonly used in DFT/MM calculations of olefin polymerization catalyzed by Ni(II) diimine complexes

shown in Fig. 2) had been shown to be efficient catalysts [39]. These compounds were mechanistically interesting because they were minor modifications of previously known complexes producing short oligomers.

A first goal of these calculations was therefore to analyze how the nature of the ligand affected the molecular weight of the resulting polymers. This involved a comparison between the energy barriers of two processes, namely the olefin insertion depicted in Fig. 1, which corresponds to chain propagation; and a β hydride transfer process like that shown in Fig. 3, which corresponds to the chain termination. Calculations were carried out on the propagation and termination steps for the two different systems shown in Fig. 2, and the results were clear. The barrier for the propagation reaction was lowered by a value between 4 and 6 kcal/mol when the bulky substituents were included, while the barrier for the termination step increased by a value between 5 and 10 kcal/mol upon introduction of the real substituents. Bulky substituents therefore facilitate the propagation of the chain, and hinder their release from the catalytic center. Both factors contribute towards obtaining longer chains, and to increasing the molecular weight of the resulting polymers, in good agreement with experiment. Moreover, examination of the optimized geometries also provided a qualitatively explanation for these results. The transition state for the propagation reaction relaxes the steric strains in the resting state by bringing the ligands together within the equatorial plane, while the transition state for the termination step increases

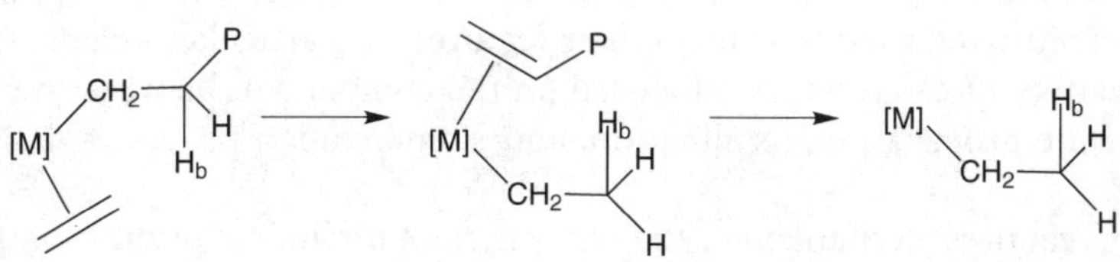

Fig. 3 A possible mechanism for the chain termination reaction in ethylene polymerization

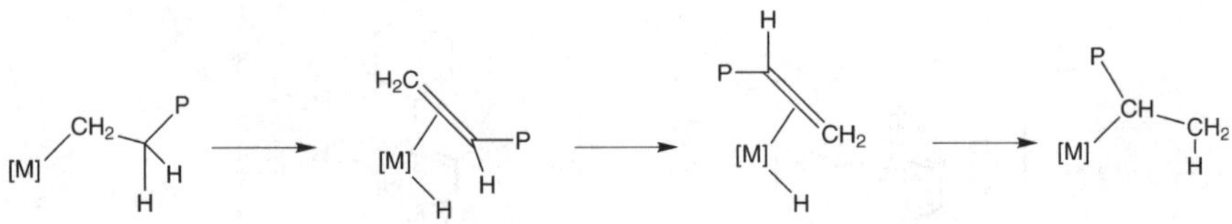

Fig. 4 A possible mechanism for the chain isomerization reaction in ethylene polymerization

the steric strain by moving the ligands away from the equatorial plane and towards the more sterically hindered axial positions.

It is worth mentioning here that the small quantitative differences found in the energy barriers computed by the two research groups are related to the different DFT and MM descriptions used in the IMOMM(DFT:MM) calculations. The particular method was IMOMM(BP86:AMBER95) in one case [36] and IMOMM(B3LYP:MM3) in the other [37, 38]. One of the studies also analyzed the catalytic efficiency of similar palladium complexes [37, 38].

A related topic that was already discussed in these first DFT/MM works is that of branching. The scheme shown in Fig. 1 would always produce always a linear polymer if ethylene was used as olefin. But a simple process of β C-H oxidative addition/reductive elimination, coupled with olefin rotation, can produce a branched polymer, as shown in Fig. 4. Calculations on the branching process for cationic diimine Ni(II) complexes [36, 37] indicated a small increase between 0.9 and 2.5 kcal/mol in the barrier for this process, associated with the introduction of the bulky substituents in the catalysts.

DFT/MM calculations on ethylene polymerization by nickel diimine complexes have been applied within Car-Parrinello molecular dynamics simulations [40, 41]. A first set of calculations was used to refine the computed energy barrier for the termination step. The enthalpy barrier computed in the calculations described above was 18.6 kcal/mol, a value which decreased to 14.8 kcal/mol at 25 °C in the molecular dynamics calculation, in better agreement with experiment [40]. A second study analyzed the capture of the olefin by the catalyst [41], and found that this process, which has no enthalpic barrier, has an entropic barrier.

Experimental research has continued to produce new and more efficient catalysts for ethylene polymerization in recent years, and computational work rapidly followed to explain their features. In particular, several articles on a variety of catalysts have focused on the comparison between the barriers for the propagation, termination, and isomerization processes described above.

The reaction mechanism of the very efficient bis(imino)pyridil complexes of iron and cobalt $[2,6\text{-}(CMe{=}N(2,6\text{-}i\text{-}Pr_2C_6H_3))_2C_5H_3N]MCH_3^+$ (M=Fe, Co) has been analyzed computationally [42–44]. The overall mechanism is similar to that depicted above for other systems, but there are some significant differences. From a theoretical point of view, the QM part of the computational study is complicated because of the existence of close energy spin states. The studies by the group of Ziegler [42, 43], based on a BP86 description, predict the reaction to take place mostly in the low spin potential energy surface (PES), singlet for iron, doublet for cobalt. In contrast to the systems described before, for these compounds the coordination of the olefin to the metal center (olefin uptake) has a barrier, and can become the rate limiting step in some of the processes. The steric bulk lowers the barrier associated with the insertion process, favoring the formation of long polymers. In fact, the barrier for the termination process becomes so high upon consideration of the steric bulk that alternative mechanisms, based on interactions with the counterion, had to be proposed. The results for the iron complex were partially questioned in a more recent work by Morokuma and coworkers [44]. In this latter work, a B3LYP description was used for the QM region. With this method, it was found that the difference in electronic energy between the different spin states is substantially smaller, and that the quintet state has a comparable energy to the singlet once the steric effects of the bulky substituents are taken into account. This quintet state is much better suited for chain propagation than for chain termination or isomerization, and the effect of the bulky substituents is therefore correlated with the stabilization of this high spin state. This set of works in fact constitutes a nice example of the importance of the QM description in any QM/MM method. DFT/MM calculations have been also used to explain the lack of inclusion of 1-hexene during polyethylene synthesis (copolymerization) when the reaction is catalyzed by Fe bis(imino)pyridil complexes [45].

A study [46] on the catalytic activity of group 4 diamide complexes $(RNCH_2CH_2CH_2NR)MCH_2CH_2CH_3^+$ (R=2,6-i-$Pr_2C_6H_3$; M=Ti, Zr) could explain the experimentally observed difference between the titanium and zirconium species. As in the case of the late transition metal complexes described above, the bulky substituents have a substantial role in increasing the termination barrier, favoring the formation of longer polymers. This effect is much larger (9.1 kcal/mol) for titanium than for zirconium (3.5 kcal/mol) because of the larger steric constraint associated with the shorter M–N distances in the first row transition metal.

More recent works of this type have not focused so much on a particular complex as on a family of complexes with small variations in the ligands. In this case, the starting point is usually a complex of modest catalytic activity, and variations of it are analyzed in order to suggest which changes must be introduced to improve efficiency. This has been done on neutral salicylaldiminato nickel (II) complexes [47] and zirconium bidentate non-cyclopentadienyl complexes [48]. The detailed reaction mechanisms are always subtle and complicated, and bulky ligands are found in most cases to play a significant role in the relative energy barriers of the reaction paths. Most of the catalysts described above are cationic, and their efficiency is usually re-

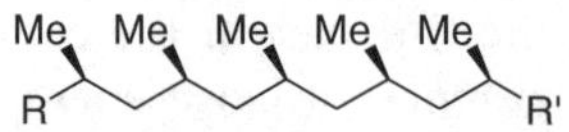

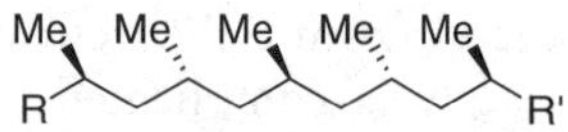

Fig. 5 Isotactic (left) and syndiotactic (right) polypropylene

lated to the nature of the counterion. The role of these counterions in the reaction mechanism is still not well understood, and research is currently being invested in this particular area [32, 49]. A recent article [50] has addressed the topic of ion-pair formation and dissociation energies between a variety of common counterions and a variety of group 4 (Cp)RTiMe$^+$ catalysts with DFT/MM methods. A full mechanistic picture is not yet available.

The second major topic in the field of olefin polymerization is that of the tacticity of the polymer [28]. If the olefin being polymerized is less symmetrical than ethylene, stereogenic centers will appear at the polymer, and the arrangement of these stereocenters can produce highly organized isotactic or syndiotactic polymers, as depicted in Fig. 5 for the case of propene polymerization. The alternative is an atactic polymer where the distribution of stereocenters is random.

From a mechanistical point of view, the introduction of a substituent into the olefin introduces two complications into the mechanism depicted in Fig. 1, which still remains valid overall. A first aspect is the regioselectivity of the insertion step; the substituent can end up in the carbon attached to the growing chain (primary insertion) or in the carbon attached to the metal (secondary insertion). A second aspect is the stereoselectivity of the insertion; the reaction can take place in either of the two non-equivalent faces of the olefin.

The topic of regioselectivity has been efficiently addressed with pure QM calculations on model systems, which have shown that it takes place mostly in a primary mode for the case of propene, although the frequency of secondary insertions is relevant because it implies the introduction of regioirregularities in the polymer chain. DFT/MM calculations on full systems have not significantly changed these conclusions.

The topic of stereoselectivity is instead intimately associated with the presence of bulky ligands. The mechanism by which the tacticity of the products is controlled in early transition metal metallocene complexes was explained mostly by theoretical calculations in the early 1990s, initially with pure MM methods. The reason why QM/MM were not used in these early stages was probably because they were not yet available in efficient implementations. The characteristic feature of these early calculations, which used different force fields throughout the years, was the freezing of an active core, which was assumed to be unchanged by the nature of the substituents and the geometry optimization of the other parameters. In some cases, the geometry was assumed from available data [51–53], while in other cases it was

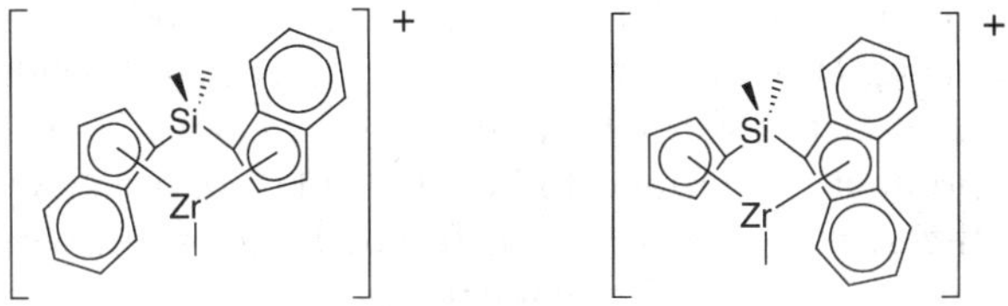

Fig. 6 Typical metallocene catalysts for olefin polymerization containing a C_2-symmetric ligand (left) and a C_s-symmetric ligand (right)

taken from ab initio calculations on model systems [54]. In a way, these calculations can be seen as a forerunner of modern DFT/MM calculations on very similar complexes. Furthermore, they are directly connected to these recent genuine DFT/MM works, and therefore merit some discussion here.

The experimental observation was that C_2-symmetric metallocene complexes of zirconium (Fig. 6) produced isotactic polymers, while C_s-symmetric metallocene complexes (Fig. 6) produced syndiotactic polymers. Pure MM calculations with frozen core showed that the stereoselectivity is not related to direct interactions of the π-ligands of the chiral metallocene with the entering monomer, but to interactions of the π-ligands with the growing chain. It is therefore the chirally oriented growing chain which discriminates between the prochiral faces of the propene monomer. For C_2-symmetric complexes, identical enantiofacial orientation in all insertion steps results in isotactic polymer formation; for C_s-symmetric complexes the enantiofacial orientation alternates between insertion steps and leads to syndiotactic polymers.

The more recent computational works on the topic of polymerization of substituted olefins have used DFT/MM methods. Since the general picture was defined with the pure MM methods described above, current research has targeted the treatment of specific aspects, and the development of new catalysts. A great deal of this research is related to variations of the catalysts depicted in Fig. 6. A usual QM/MM partition is shown in Fig. 7. It is clear from this that the MM region is crucial, because otherwise the C_2- and C_s-symmetric catalysts would be identical.

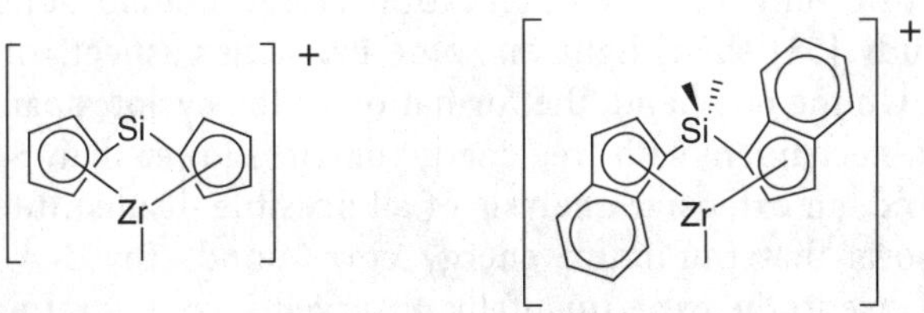

Fig. 7 DFT model (left) and full DFT/MM (right) descriptions commonly used in DFT/MM calculations of olefin polymerization catalyzed by metallocene complexes

The behavior of rac-$Me_2C(3\text{-}t\text{-}Bu\text{-}1\text{-}Ind)_2ZrCl_2$ in propene polymerization was analyzed using DFT/MM methods [55]. This catalyst produces a highly regiospecific and fairly isospecific polymer. The energy barriers leading to the isotactic product were found to be 4.0–6.0 kcal/mol below those corresponding to the syndiotactic product, in good agreement with the experimental value of 4.6 kcal/mol. Mechanistic aspects related to the presence of irregularities in the resulting polypropylene were also analyzed. The relatively high rate of chain-end epimerization, a process by which the methine group of the last inserted unit is inverted, could be rationalized by the relative stability of the product of the β-H transfer to the metal. Allylic activation of the growing chain end followed by propene insertion was found to be a possible source of internal vinyledene unsaturations, but probably not of epimerization. The differences to the less sterically constrained rac-$C_2H_4(1\text{-}Ind)_2ZrCl_2$ catalyst could therefore be rationalized.

A joint experimental and computational DFT/MM study [56, 57] on the copolymerization of ethene and 2-butene provided further proof of the validity of the growing chain control of stereoselectivity mentioned above. The idea is that, if the same steric interactions postulated for propene hold for 2-butene, insertion of *Z*-butene should be favored with C_2-symmetric metallocenes like $Zr(Me_2Si(1\text{-indenyl})_2CH_3^+$, while *E*-butene should be favored with C_s-symmetric metallocenes like $Zr(Me_2Si(\text{cyclopentadienyl-9-fluorenyl})CH_3^+$. DFT/MM calculations confirmed this qualitative view, predicting the barrier for *Z*-butene to be 1.6 kcal/mol lower in the case of C_2-symmetric complexes, and the barrier for *E*-butene to be 1.8 kcal/mol lower in the case of C_s-symmetric complexes. These results were corroborated by experiments, which showed molar compositions of 14% and 25% when the appropriate 2-butene isomer was copolymerized with ethene, while the molar percent was in the range of 1% when the wrong isomer was used.

The C_2-symmetric rac-$H_2C(3\text{-}t\text{-}Bu\text{-}1Ind)_2ZrCl_2$ catalyst shows remarkable behavior during the copolymerization of ethylene with 1,3-butadiene [58]. In most experimental conditions, it produces ethylene copolymers containing previously unobserved methylene-1,2-cyclopropane together with methylene-1,2-cyclopentane units. In special conditions of high temperature and low ethene concentration, the products are ethene copolymers containing previously unobserved 1,1- and 1,3- constitutional units from 1,3-butadiene, both presenting only the *E* configuration of the double bond. A detailed DFT/MM study [58] sheds light on some particular aspects of the reaction mechanism. On the one hand, the formation of the cyclopropane units is explained by a mechanism with free energy barriers lower than 5 kcal/mol. On the other hand, an extensive analysis of all possible disubstituted allyl intermediates shows that minimum energy corresponds (by 3–4 kcal/mol) to those giving rise to the experimentally observed *E* configuration of the double bond.

A final application of DFT/MM calculations to olefin polymerization by metallocene catalysts comes from oscillating metallocene catalysts [59]. These are remarkable complexes of general formula $[(2\text{-Ar-indenyl})_2MP]^+$ (M=Zr, Hf; P=polymeril), very similar to the usual metallocene catalysts (Fig. 6), the most notable difference being the lack of the bridge connecting the two metallocene rings. These particular catalysts produce a peculiar polypropylene which contains isotactic and atactic sequences. It has recently been suggested that the isotactic sequences are associated with the presence of a conformationally locked catalytic species with restricted ring mobility. It is proposed that this stereorigidity is induced by the proximity of a counteranion, which would play the role of the interannular bridge of the metallocene complexes described above (Fig. 6). A DFT/MM calculation of the $[rac\text{-}(2\text{-phenylindenyl})_2ZrMe(C_3H_6)][B(C_6F_5)_4]$ ion couple is presented to support this hypothesis, with the ion pair formation energy being compatible. However, this work falls short of a full characterization of the mechanism. As mentioned above already [50], the role of the counteranion in olefin polymerization is still a topic of current active research, and a full understanding has not been yet achieved.

Two communications on propene polymerization by non-metallocene catalysts that include DFT/MM calculations have been recently published [60, 61]. They deal with group 4 bidentate non-cyclopentadienyl complexes. In the first communication [60], the topic addressed is the fact that a C_2-symmetric precatalyst of titanium leads to a syndiotactic polymer, contrary to observations of metallocene catalysts. The chirality at the metal center is found to play a key role in the stereocontrol of the process. The second communication [61] addresses the fact that a C_2-symmetric precatalyst of zirconium very similar to the previous one produces an isotactic polymer, finds out that it is due to a complicated concourse of synergic steric and electronic effects, and emphasizes the key role that serendipity still plays in the design of new catalysts.

As shown in this section, olefin polymerization is an area where DFT/MM calculations have found a large number of applications. The first works, some of them using pure MM methods, dealt with fundamental mechanistic aspects, while more recent ones deal with specific topics that are being currently investigated by experimental groups, in many cases in direct collaboration.

3 Hydrogenation of Alkenes

The development of enantioselective homogeneous catalysis of alkenes represents an outstanding achievement in modern chemistry, as recently recognized by the Nobel Chemistry award committee, when they gave the prize to

Profs. Noyori and Knowles in 2001. The Wilkinson catalyst was the first well-characterized homogeneous transition metal catalyst for alkene hydrogenation [62]. The asymmetric version of the hydrogenation was mainly introduced by the groups of Knowles [63] and Büthe [64] using chiral phosphines, and by the group of Kagan [65] using chelating diphosphine ligands. Further development of asymmetric hydrogenation (Noyori, Burk, and so on) [66, 67] has created a robust and technologically important set of catalytic asymmetric synthetic techniques. Among the most popular processes is the well-known Monsanto's L-Dopa synthesis [68].

In the general hydrogenation mechanism, the initial step is the solvent coordination to the metal center, followed by H_2 addition to the catalyst to form the dihydride complex. In a subsequent step, the alkene binds to the catalyst and is then transformed into the product via migratory insertion and reductive elimination steps. Such a mechanism is called the "hydride pathway". An alternative mechanism not involving stable hydride complexes takes place when chelating phosphines are bound to the catalyst. In this case, the hydrogenation goes through a reversible alkene binding followed by addition of dihydrogen. This sequence is termed the "alkene pathway".

Experimental studies by the groups of Halpern and Brown [69, 70] on the hydrogenation of prochiral alkenes by a chelating-diphosphine catalyst provided a detailed, and unexpected, picture of the overall mechanism of hydrogenation. The reaction of the catalyst with a prochiral alkene can produce two diastereomeric catalyst-alkene complexes. The most stable (and most abundant) diastereomeric complex is relatively unreactive, whereas the small amount of the complex in the less populated diasteromeric state gives the productive catalytic cycle. Landis has coined the phrase "anti-lock-and-key" motif to describe this mechanistic model.

In 1993 Burk established that DuPHOS (the chiral chelating phosphine developed at Dupont) complexes exhibit the same "anti-lock-and-key" mechanistic motif as seen for the aryl phosphine ligated catalyst [71]. Recently, Imamoto's mechanistic investigations of their electron-rich disphosphine catalyst pointed to many of the same features seen in the aryl diphosphine and DuPHOS systems, such as anti-lock-and-key behavior and complete reversal enantioselectivity with an electron donating substituent [72]. However, they also note that solvated catalyst compounds *can* react with hydrogen prior to forming the catalyst-alkene adduct. They observed that the dihydride pathway becomes more favorable as the phosphine ligand bite angle increases.

Experimental investigations by Halpern and Landis on the hydrogenation of methyl-(*Z*)-α-acetamidocinnamate by Rh(DiPAMP) allowed them to propose the reaction mechanism discussed above. These studies also showed that minor changes on the substrate can drastically modify the enantiomeric excess. They found that enantiomeric excess depends on: (i) the nature of the alkene substituents, and (ii) their relative positions (*E* or *Z*); the best re-

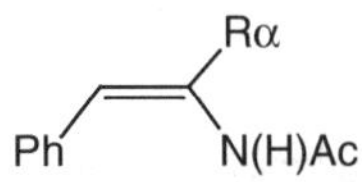

Fig. 8 Optimal arrangement of the enamide substituents for catalytic hydrogenation

sults are obtained when the Rα substituent is an electron-withdrawing group, placed Z to the amide and with Ph in the β position (Fig. 8).

With the aim of giving an explanation for these experimentally observed phenomena, the Landis group began a theoretical investigation of the mechanism for the hydrogenation of α-formamidoacrilonitrile by [Rh(R,R)-Me-(DuPHOS)]$^+$. The initial work was on model systems using DFT methods [73]. After that, they applied the ONIOM method [12] on the complete real system, therefore introducing the chirality issues into their investigations [74].

Their initial research was a QM study at the B3LYP level on a model system, where they chose [Rh(PH$_3$)$_2$(α-formamidoacrylonitrile)]$^+$ as a model catalyst-alkene complex [75]. Since this model system is achiral, they do not need to consider different diastereomeric manifolds to study the reaction mechanism. Nevertheless, they need to consider four different reaction pathways corresponding to the four possible *cis*-dihydride isomers that can be formed on the addition of H_2 to the metal complex (Fig. 9).

From analysing these four reaction pathways, the authors propose pathway A as the operative one. The reaction begins with the formation of a van der Waals complex between the rhodium complex and molecular hydrogen. This van der Waals complex is found in other catalytic cycles [76], and has been predicted by other theoretical studies [77]. After that, the structure of

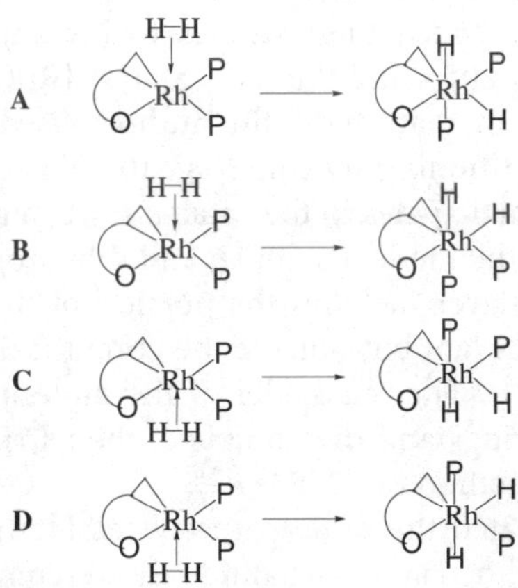

Fig. 9 Isomers corresponding to the formation of dihydride complex from the addition of H_2 to [Rh(PH$_3$)$_2$(α-formamidoacrylonitrile)]$^+$

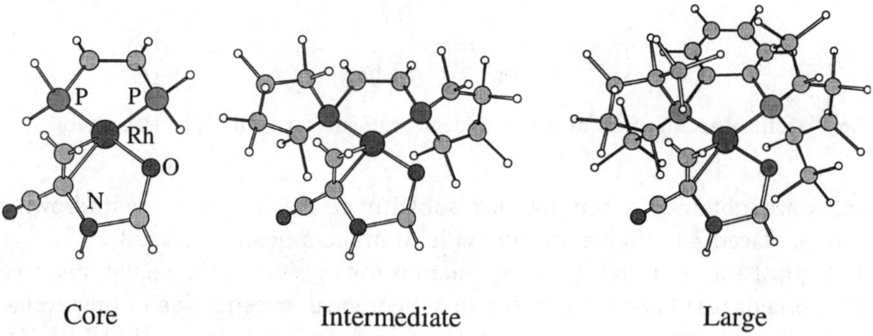

Fig. 10 Partitioning scheme for ONIOM calculations of the "[Rh((R,R)-Me-DuPHOS)]$^+$ +enamide" system

the metal complex changes from square planar to trigonal bipyramidal, with the H_2 occupying a coordination site. Next step is the oxidative addition of the dihydrogen to form the hexacoordinated dihydride intermediate. The next two steps correspond to the migratory insertion of the carbons of the double bond to the Rh-H bonds. In the lowest energy mechanism overall, this insertion goes through the migration of the β-carbon into a Rh–H bond to form an α-alkyl hydride. Finally, in the last step reductive elimination takes place, forming the hydrogenated product. This last step has the largest barrier of the overall mechanism.

This mechanism differs from the commonly accepted mechanism of catalytic asymmetric hydrogenation, because the accepted one proposed an irreversible oxidative addition (the oxidative addition step should have the largest energy barrier), but it does agree with recent experimental results that show irreversible and stereodetermining migratory insertion [78].

Once the reaction mechanism for the studied model system was established, the authors extended their studies to investigate the origin of the enantioselectivity [79, 80] using the real catalyst [Rh((R,R)-Me-DuPHOS)]$^+$ in their calculations. In that study, the authors used the ONIOM method with a three-layer partitioning scheme (Fig. 10). The core layer contains the region that actively participates in the catalysis: Rh, a bridging bis(phospholano)ethylene ligand, the model enamide, and dihydrogen (at the B3LYP level). The intermediate layer includes the portion of the molecule that is the most electronically relevant but outside the core: the 5-membered phospholane rings (at HF level). The remainder of the molecule (the methyl groups on the phospholane rings and the complete phenyl ring of the bridge) was treated with an MM method (at UFF level).

The authors calculated the same pathways as for the model system, but starting from both of the catalyst-enamide diastereomeric forms previously determined. Their results show that the mechanism goes through the same reaction pathway as for the model system (pathway A) but with a significant difference: the migratory insertion barriers are noticeably smaller, particularly in relation to the oxidative addition.

Their calculations reproduce the experimental observations of the anti-lock-and-key behavior and the formation of the *R* enantiomer as the major product. In the initial step, the catalyst-enamide complex with the enamide bonded to the catalyst through the pro-*R* face is 3.6 kcal/mol less stable than the corresponding diastereomer with the enamide bonded through the pro-*S* face. Nevertheless, the overall energetic pathway corresponding to the formation of the *R* product is 4.4 kcal/mol more stable than the one corresponding to the *S* product. Therefore, the difference in energy of the initial diastereomers corresponds to a solution concentration of the pro-*R* complex of less than 0.2%, whereas the difference in the overall energetic barriers corresponds to an enantiomeric excess of 99.9% for the *R* product. These results are in excellent agreement with experiment and are the explanation for the anti-lock-and-key behavior.

The study was then broadened to examine the intriguing effect of the Rα substituent on the enamide, since it can completely change the reaction enantioselectivity [81]. In that study, the nitrile group was replaced by a *t*-Bu group to generate *N*-(1-*tert*-butyl-vinyl)-formamide. The calculations were carried out as in the previous system, replacing the nitrile group by the *t*-Bu group. The *t*-Bu group was partitioned in two ONIOM layers, placing the full group in the intermediate layer (HF) with a methyl group in the core layer (B3LYP).

The optimized structures of the diastereomeric catalyst-enamide complexes with the enamide bonded from its pro-*R* and pro-*S* faces have very similar structures to their nitrile counterparts. The calculated free energy difference between these two diastereomers is practically zero (0.07 kcal/mol).

The pathway with the lowest overall barrier, pathway C, is in this case the one starting from the pro-*S* diastereomer producing the *S* enantiomer, in accord with experiment. The overall energetic difference between the lowest pathways giving the *S* product and the *R* products is 6.5 kcal/mol, which corresponds to an enantiomeric excess >99% that is also in excellent agreement with the experimental results.

Comparing the reaction mechanisms for both Rα substituents, the hydrogenation of *N*-(1-*tert*-butyl-vinyl)-formamide goes through pathway C, giving the *S* product; this mechanism implies that the H_2 molecule adds to the catalyst-enamide complex from the "bottom side" (see Fig. 9). Hydrogenation of α-formamidoacrylonitrile, however, gives the *R* product; this mechanism goes through pathway A, which implies that the H_2 molecule adds from the "top side". The addition of H_2 to the complex implies movement of the ligands around the metal center. In this motion, the alkene goes to a less hindered position. Hence, depending on the alkene substituents, the less hindered position is reached by pathway A (CN) or pathway C (*t*-Bu).

The next step, the migratory insertion step, can produce two different hydrides: an α-alkyl hydride, or a β-alkyl hydride, depending on the carbon of the double bond undergoing the insertion. The β-alkyl hydride of the nitrile-substituted enamide is considerably less stable than the α-alkyl hydride (>13 kcal/mol), which also implies larger migratory insertion barriers to form the β-alkyl hydride, where greater atomic motion is needed. In the other case, for *t*-Bu substituted enamide the β-alkyl hydride is more stable than the α-alkyl hydride, also implying a lower energetic barrier with less atomic motion. This shift is attributed to the destabilizing effect that an electron-donating group has on a metal-carbon bond, which strongly affects the stability of the α-alkyl hydride.

The computational study of the model system at QM level, followed by the ONIOM study on the real system, was able to answer some key questions about the asymmetric hydrogenation of alkenes. These theoretical investigations helped us to understand the "anti-lock-and-key" motif and the origin of the intriguing enantioselectivity upon changing the substituent of the alkene from nitrile to *t*-Bu.

4 Osmium Dihydroxylation of Olefins

Osmium-catalysed dihydroxylation of olefins is a powerful route towards enantioselective introduction of chiral centers into organic substrates [82]. Its importance is remarkable because of its common use in organic and natural product synthesis, due to its ability to introduce two vicinal functional groups into hydrocarbons with no functional groups [83]. Prof. Sharpless received the 2001 Nobel Prize in chemistry for his development of asymmetric catalytic oxidation reactions of alkenes, including his outstanding achievements in the osmium asymmetric dihydroxylation of olefins.

The fundamental mechanism of the dihydroxylation of olefins remained unclear for a long time. Two different mechanisms, concerted [3+2] and stepwise [2+2] were popular, plausible, mechanisms for the reaction [84, 85]. The controversy about the reaction mechanism was finally resolved after the thorough work of several experimental [84, 85] and theoretical [86] research groups. A scientific consensus emerged considering the [3+2] mechanism as the operative one [87]. Nevertheless, in very special cases and using other metal oxides, the activation barrier for the stepwise mechanism is lower in energy than for the concerted one [88, 89]. Other mechanisms, such as a diradical mechanism, have been also discussed [90].

The preference for the [3+2] mechanism does not provide in itself an explanation for the high enantioselectivity observed in these reactions. The theoretical studies undertaken to discern the mechanism were carried out at quantum mechanical level on a model achiral system. To consider the asymmetry in the reaction, the calculations required the introduction of a chiral amine ligand (such as cinchona alkaloid) coordinated to the osmium and a prochiral alkene.

With this aim, the group of Norrby developed a transition state force field for the study of the asymmetric dihydroxylation reaction [91]. This force field is purely developed from quantum mechanical reference data [92]. In their studies they use different ligands from the first generation (where the amine ligands are the alkaloids dihydroquine or dihydroquinidine) and second generation (where a symmetric linker couples two alkaloid units), and several alkenes. The calculated ee's are in very good agreement with experiment.

The group of Chapleur has developed a computational tool for predicting the stereochemical outcome of asymmetric reactions, and has applied the method to osmium asymmetric dihydroxylations [93]. The method is related to those used in flexible docking, and is based on molecular mechanics force-fields. It consists of applying a genetic algorithm to carry out a conformational search, using a free energy function as a scoring function, similar to that used in docking programs. Despite the simplicity of the method, its application to a variety of olefins gives good results compared to experiment, providing a semiquantitative method to predict the enantioselectivity of this reaction.

Our group has carried out a quantitative theoretical characterization of the origin of enantioselectivity in the dihydroxylation reaction of olefins by means of the hybrid QM/MM IMOMM(B3LYP:MM3) method. Following our previous studies on this reaction [94], we studied the origin of the enantioselectivity for olefins with aromatic and n-aliphatic olefins. In these investigations we used a real catalyst, $OsO_4{\cdot}(DHQD)_2PYDZ$. The partitioning of the system into the QM and MM regions was done as shown in Fig. 11. In the catalyst, the $OsO_4{\cdot}NH_3$ moiety was described at B3LYP level, whereas the rest of the amine ligand was described at MM3 level. As far as the olefin is concerned, different QM regions were adopted depending on the olefin class. In the study of aromatic olefins, $CH_2{=}CH_2$ was described at B3LYP level, whereas the Ph substituent was described at MM3 level. In the case of aliphatic olefins, $CH_2{=}CH{-}CH_3$ was described at B3LYP level, whereas the rest of the n-aliphatic chain (from propene to 1-decene) was described at MM3 level.

Once the [3+2] mechanism was accepted as the operative one, to investigate the transition states associated with the formation of the osmate ester in a chiral system, one must take into account all of the different ways that an olefin can approach the catalyst. These different paths were classified according to the criteria shown in Fig. 12. The olefin binds to an axial and to an equatorial oxygen, providing three different families of reaction paths labeled as "regions" **A**, **B** and **C**. When the olefin has one substituent, it can be placed in four different "orientations", which are labeled as **I**, **II**, **III**, **IV**. The

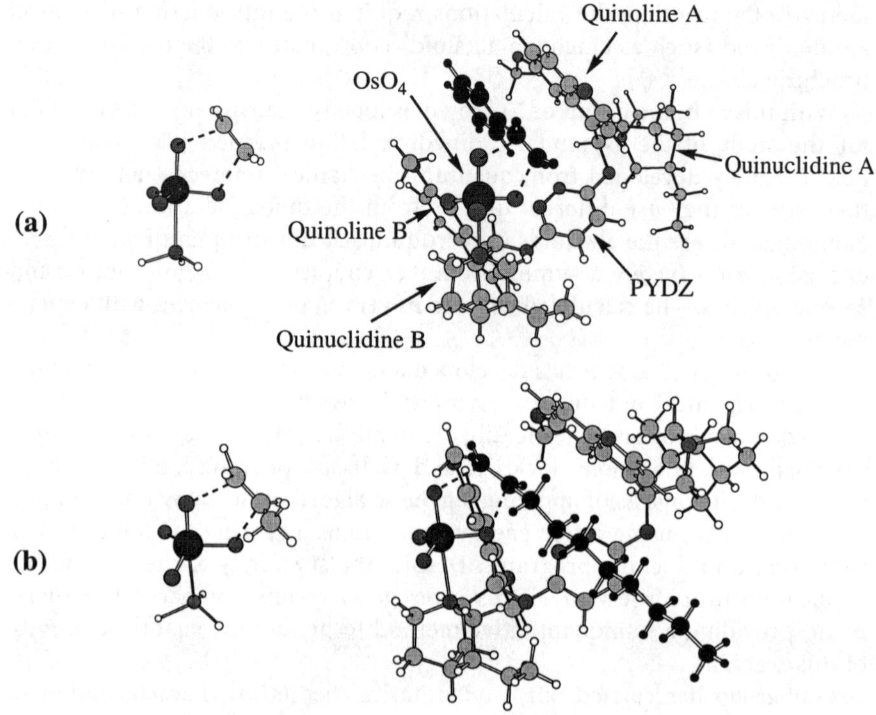

Fig. 11 Partitioning scheme for IMOMM calculations of the dihydroxylation of **a** styrene and **b** 1-decene by $OsO_4 \cdot (DHQD)_2PYDZ$. Alkenes highlighted in black

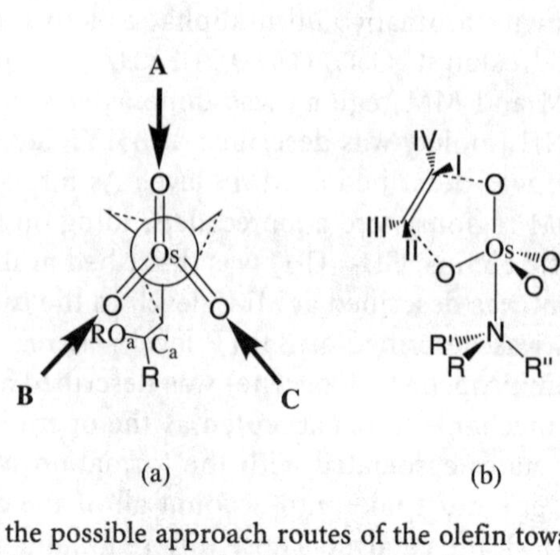

Fig. 12 Definition of the possible approach routes of the olefin toward the catalyst. **a** Top view along the O-Os-N axis showing the three different "regions" of entry of the olefin (A, B, C). **b** Side view perpendicular to the O-Os-N axis showing the four possible "orientations" of the olefin substituent (I, II, III, IV)

joint consideration of the three regions and the four possible positions of the olefin substituent yield a total of twelve different pathways.

As a representative example of the asymmetric dihydroxylation of aromatic alkenes, we studied the asymmetric dihydroxylation of styrene [95]. The twelve transition states corresponding to all of the possible regions and orientations of entry of the styrene were characterized. As far as the region of entry is concerned, the three lower energy saddle points correspond to region **B**: **B-I**, **B-III** and **B-IV**, with relative energies of 0.0, 0.1 and 2.6 kcal/mol, respectively. Therefore, there is no doubt that computationally **B** is the preferred region of entry for the olefin, in complete agreement with experimental observations obtained from kinetic studies [96]. The fact that region **B** is preferred over the other regions is given by the nature of the steric interactions between the substrate and the catalyst. Region **A** is the least hindered region, so the magnitude of steric interactions should be smaller in this region than in region **B**. The fact that the energies of the saddle points are lower in region **B** can only mean that the steric interactions are of an attractive nature. This is fully consistent with the existence of an intermediate in the reaction path previously characterized [94b].

After showing that the reaction goes through region **B**, the authors analyzed the different orientations of the olefin within this region, since the enantioselectivity is decided by the orientation of the olefin with respect to the catalyst. Isomers **I** and **III** lead to the *R* product, whereas isomers **II** and **IV** lead to the *S* product. **B-I** and **B-III** are the lowest energy saddle points leading to the *R* product, with relative energies of 0.0 and 0.1 kcal/mol, respectively. The lowest energy saddle point leading to the *S* product is **B-IV**, with a relative energy of 2.6 kcal/mol. The computed enantiomeric excess using these energetic values is 99.9%, in good agreement with the experimental value of 96% [84].

The properties of these transition states were analyzed in some detail, since this is the point where the selectivity is decided. Comparing the relative QM and MM energies for each of the saddle points, it was found that the major differences were in the MM part, with QM energies remaining very similar. With the aim of analyzing the origin of the difference in the MM term, the interaction energies between the catalyst and the substrate were evaluated for each of the transition states. The catalyst was divided into 5 parts: Quinoline A, Quinoline B, PYDZ, OsO_4, and the rest (including both of the remaining quinuclidines). Both quinolines and the PYDZ account for more than 75% of the interaction energy between the catalyst and the substrate, divided up as follows: Quinoline A 53%, Quinoline B 20% and PYDZ 12%. Therefore, although Quinoline B and PYDZ contributions are significant, Quinoline A plays a key role in deciding the stereochemistry of this reaction. The main interactions between the substrate and the catalyst are π-π interactions, face-to-face with the quinolines, and edge-to-face with the PYDZ.

We have recently broadened those investigations to study the origin of the enantioselectivity in the dihydroxylation of terminal aliphatic n-alkenes. The dihydroxylation of the series from propene to 1-decene was studied by means of the IMOMM method [97]. Experimental studies on propene, 1-butene, 1-pentene, 1-hexene and 1-decene showed that the reaction was enantioselective in all cases, leading to the *R* product. Moreover, the results show a dependence of the enantioselectivity on the chain length; it sharply increases from propene to 1-pentene, and after that the enantioselectivity remains practically constant for 1-hexene and 1-decene. The explanation for this dependence of the enantioselectivity with the chain length remained elusive. On the other hand, the π-stacking interactions that were found to be critical for styrene cannot be responsible for the observed enantioselectivity for these terminal aliphatic n-alkenes because they do not have aromatic rings.

The computational study of the osmium dihydroxylation of aliphatic alkenes is much more complicated than the case of aromatic alkenes due to the large number of conformations that the former could adopt. To overcome this issue, we considered the system to be composed of two different parts: the catalyst and the olefin. For the catalyst, the conformation considered is that from the X-ray structure. As already shown in the study of styrene [95], and in some experimental works [98], the catalyst is a fairly rigid molecule. For the aliphatic alkenes under study, there is a large number of possible conformations; in addition, the stability of an olefin conformation is also affected by the interactions between the olefin substituent and the catalyst. Therefore, the catalyst must be included in the conformational search. The conformational analysis was done using a scheme based on the "systematic search" approach [99]. The strategy consisted of two parts: first we developed a method to identify all of the possible conformations; afterwards, we screened all of the possible conformations at MM level to select the most stable. Finally, we only carried out the relatively expensive QM/MM calculations on these selected conformations.

The "catalyst + aliphatic n-alkene" system, for all of the olefins from propene to 1-decene, give almost 40,000 possible conformations. The most stable conformations for each of the olefins were selected (around 1,700) and their transition states were optimized at IMOMM level. The calculated enantiomeric excesses are shown in Fig. 13. Calculations are able to reproduce the observed increase in ee for short chains, and the presence of a ceiling value after which the increase in enantioselectivity is much smaller, in excellent agreement with experiment.

Once the experimental trends were properly reproduced by the calculations, we analyzed the main differences between the *R* and *S* enantiomers in order to identify the origin of the enantioselectivity for these olefins and its dependence on the chain length. The part of the olefin that has the greatest interaction with the olefins is Quinoline A, though its behavior is very similar for both enantiomers; therefore, although Quinoline A is decisive for cat-

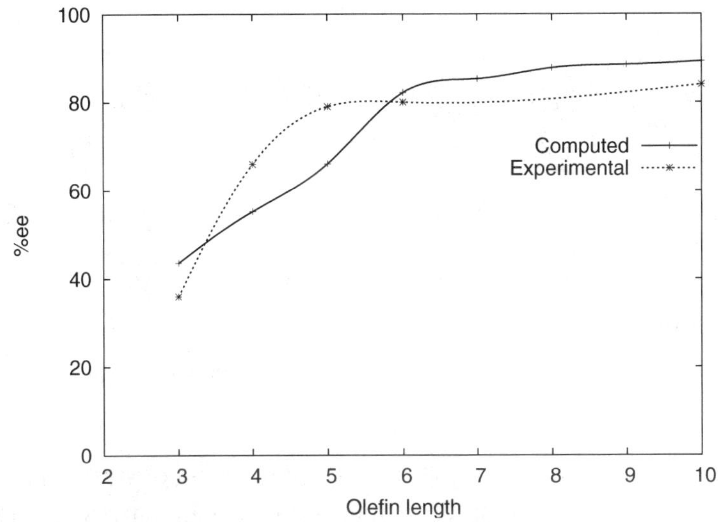

Fig. 13 Experimental and computed enantiomeric excesses as a function of the chain length of the aliphatic n-alkene

alytic purposes, is not responsible neither for enantioselectivity nor for its dependence on the chain length. There appears to be a significant difference in the way the catalyst interacts with aromatic and aliphatic alkenes. The second largest contribution to the interaction energy corresponds to the spacer PYDZ, with different behavior for the *R* and *S* enantiomers. This is noteworthy because it means that the PYDZ region is one of the key factors that decides the enantioselectivity. The interaction energies of Quinuclidine B (see Fig. 11) with the substrate show similar behavior to PYDZ for the *R* and *S* enantiomers, although their absolute values are smaller. The interactions between the olefins and the remaining parts of the catalyst, Quinoline B, OsO_4 and Quinuclidine A are quite small in all cases compared to those just commented. As a result, the enantioselectivity of dihydroxylations of aliphatic n-alkenes is determined by the interactions of the olefin with two regions of the catalyst: PYDZ and Quinuclidine B.

Figure 14 shows the structures of the most stable transition states leading to the *R* products, for propene, 1-hexene and 1-decene. The authors claim that one can see how the elongation of the olefin chain from propene to 1-hexene increases the interactions between the olefin and the three most important parts of the catalysts, Quinoline A, PYDZ and Quinuclidine B. Moving from propene to 1-hexene, the carbon atoms added to the aliphatic chain are still close to these three parts of the catalyst, and can have strong interactions. In contrast, once the aliphatic n-alkene has six carbon atoms, all of the carbon atoms subsequently added to the chain are too far from the catalyst to interact strongly with it. These interactions between the aliphatic

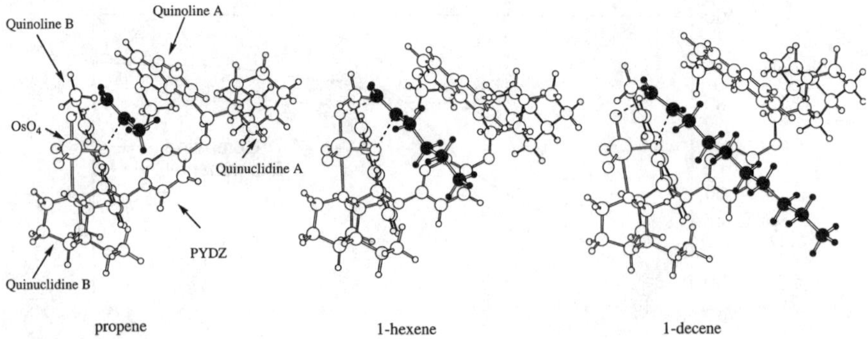

Fig. 14 Views of the selected transition states giving the *R* product for: propene, 1-hexene and 1-decene. The olefin is highlighted in black

chain of the alkene and the aromatic rings of the catalyst (PYDZ and Quinoline A) can be described as CH-π interactions [100], whereas the interactions with Quinuclidine B can be described as hydrophobic-hydrophobic interactions [101].

The origin of enantioselectivity in the asymmetric dihydroxylation of aromatic and aliphatic alkenes catalyzed by OsO_4·$(DHQD)_2$PYDZ has been analyzed by means of theoretical calculations that use the IMOMM method. The enantiomeric excesses computed for both types of olefins are in excellent agreement with experiment. Moreover, the origin of the enantioselectivity was also identified, being of π-stacking nature for aromatic olefins, and CH-π plus hydrophobic interactions for aliphatic terminal n-alkenes. The most relevant parts of the catalyst responsible for enantiomeric differentiation have been also recognized.

5 Hydroformylation by Rhodium Catalysts

The hydroformylation of olefins is one of the largest and most prominent industrial catalytic processes, producing millions of tons of aldehydes annually [102]. Initially, cobalt-carbonyl species were used as catalyst, though rhodium complexes modified by special ligands, usually phosphines, are predominantly used nowadays. Over the last two decades, continued development of new phosphine and phosphite ligands has allowed significant advances in hydroformylation chemistry, especially with respect to catalyst selectivity and stability [103].

The widely accepted mechanism for olefin hydroformylation using a $HRh(PR_3)_2(CO)$ catalyst system was proposed over 30 years ago by Wilkinson et al [104]. The catalytic cycle comprises many of the fundamental reactions of organometallic catalysis, and is described as follows: (a) olefin coordination; (b) olefin insertion; (c) CO insertion; (d) oxidative addition of H_2, and; (e) reductive elimination of aldehyde. Kinetic investigations suggest the rate-determining steps as (d) and (e), respectively. Step (b), in turn, is fast and gives rise to the formation of linear and branched aldehydes. For that reason, olefin insertion is the crucial step in determining the regioselectivity.

Initial theoretical investigations of the mechanism were at QM level on model systems in the late 1980s [105]. Afterwards, Ziegler and coworkers studied the complete [106] hydroformylation cycle using cobalt-carbonyl species by means of DFT methods. Frenking's group [107] has explored a number of elementary steps in $HRh(CO)_4$ catalyzed hydroformylation. Morokuma and coworkers applied ab initio calculations (HF and MP2 level) to $HRh(PH_3)(CO)_2$ complexes [108], whereas the groups of Cundari [109] and de Almeida [110] carried out studies on related $HRh(PH_3)_2(CO)$ model complexes at DFT and MP4 levels, respectively. Pure MM methods have also been employed to investigate ligand-substrate interactions in intermediates of the hydroformylation process, and their influence on regioselectivity [111]. Nevertheless, despite these theoretical investigations and the wide variety of phosphine and phosphite ligands studied experimentally, there was no detailed understanding of how these phosphines control the regiochemistry.

In order to solve this question, Herrmann and coworkers employed a combination of QM and MM methods to study the experimental phosphines computationally [112]. They performed DFT calculations on model rhodium complexes, and subsequently, based on the frozen geometries of the transition states, force-field calculations were carried out in order to incorporate steric effects. Their studies concentrate on the step where the regioselectivity is defined, the olefin insertion [113], calculating their transition states on complexes with two equatorially coordinated phosphine ligands. Using the relative transition state energies calculated with this combined QM and MM method, they reproduce the regioselectivity tendencies in the hydroformylation of ethylene and propene by several selected phosphine ligands (DIPHOS, BISBI, NAPHOS and TPP).

In other works, the same authors used this combined QM and MM method to rationalize the stereoselective hydroformylation induced by disphosphine ligands. Firstly, they elucidated the principle that accounts for the outstanding properties of the C_1-symmetric phosphine-phosphite ligand BINAPHOS [112b]. In a subsequent study, Gleich and Herrmann focus on another six C_2-symmetric diphosphine ligands [114]. They state that the performance of these ligands in asymmetric hydroformylation is mainly governed by two interdependencies: (i) the smaller the chelate ring, the more diminished its influence on asymmetric induction for the only available axial-equatorial coordination modes; (ii) a larger chelate ring often entails a greater backbone flexibility, and the equatorial-equatorial coordination mode must also be considered.

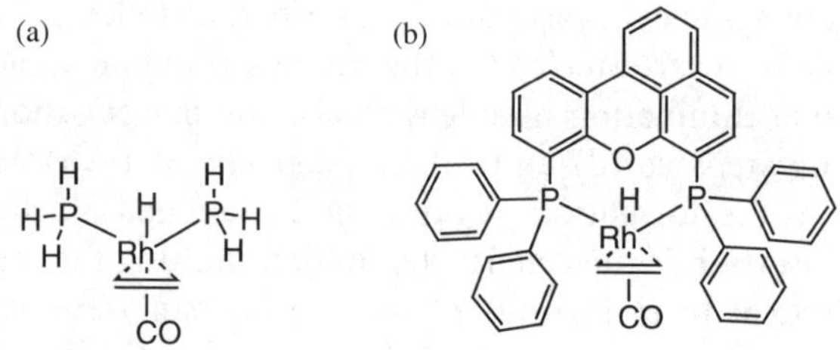

Fig. 15 Schematic representation of the atoms included in the QM **a** and MM **b** parts of the QM/MM calculations on the hydroformylation of propene by RhH(CO)(benzoxantphos) catalyst

The origin of regioselectivity in rhodium diphosphine catalyzed hydroformylation was first studied by means of a coupled QM/MM method, the IMOMM method, by Carbó et al [115, 116]. The authors selected the hydroformylation of ethene and propene by RhH(CO)(diphosphine) for their studies, where the diphosphine ligands were benzoxantphos and homoxantphos. They selected these phosphine ligands to gain insight into the relationship between the bite angle and the regioselectivity. The alkenes and the $RhH(CO)(PH_3)_2$ complexes were described at QM level, whereas the rest of the system (the xantphos ligands) was described at MM level (see Fig. 15).

In their investigations, the authors also assumed that the regioselectivity is defined in the olefin insertion step. Moreover, based on previous ab initio calculations in model systems and experimental measurements, the authors considered that the key intermediate has the phosphine chelating ligand occupying an equatorial-equatorial position. This intermediate can have two different isomers, depending on the apical positions occupied by the CO and the H ligands in the trigonal bipyramidal coordination.

To determine the relationship between the structure and activity of this intermediate, the four different conformations that it can adopt upon the coordination of the terminal alkene (propene) were investigated. Moreover, each of these conformations can generate two products, depending on how it rotates the double bond, clockwise or counterclockwise. Therefore, there are eight possible pathways for olefin insertion to the Rh–H bond, and consequently, eight possible transition states. All of them were located, and the regioselectivity was calculated using the relative stabilities of all of these transition states, assuming a Boltzmann distribution.

The results showed that the transition states giving a linear product are lower in energy than those giving the branched product, with calculated percentages for propene linear product over branched of 83% and 73% for benzo and homoxantphos, respectively. Computed values are in good agreement with the reported experimental values of 98.1% and 89.5%, respectively. As far as geometries are concerned, the authors did not notice any relevant difference, apart from different ranges in the bite angle for the two systems, ranging from 111.0 to 112.3 for benzoxantphos and from 99.5 to 104.3 for homoxantphos. Therefore, the more selective diphosphine, benzoxantphos, has a range of larger bite angles than the less selective homoxantphos. The observed experimental increase in regioselectivity for linear aldehydes with increasing natural bite angle was therefore reproduced.

The next target was to identify the specific role played by the bite angle, and two possibilities were considered. One of them was an orbital effect, where the arrangement of the phosphines could change the electronic distribution at the metal and so affect its reactivity. The second possibility was a pure steric effect, with the larger bite angles pushing the ligands towards the substrate and so affecting the selectivity. This problem was investigated via a computational experiment on a hypothetical system obtained by replacing each phenyl substituent by hydrogen and maintaining the backbone of diphosphine ligands. This hypothetic system was labeled the "PH_2 model". The results for these PH_2 model systems showed, like in the real system, a different range of bite angle values for the two types of phosphine ligands. The bite angle of model complexes increases around 7° for benzoxantphos, and around 2° for homoxantphos, respectively. In contrast, the selectivity diminishes. Computed regioselectivities in the PH_2 model system for linear product over branched are 74% for benzoxantphos and 63% for homoxantphos, whereas for real diphosphines they are 83% for benzoxantphos and 73% for homoxantphos, respectively. The bite angle does not have a direct orbital effect on regioselectivity, because the larger bite angles of the PH_2 catalysts bring lower selectivities. The effect of the bite angle is associated with the purely steric influences of the phenyl substituents of the phosphine ligands, which are pushed closer to the substrate as the angle increases.

Shortly after, Decker and Cundari carried out a computational investigation of propene insertion into the Rh–H bond of $HRh(PPh_3)_2(CO)(\eta^2$-$CH_2{=}CHCH_3)$ using the two-layered ONIOM(B3LYP:UFF) approach [117]. The $HRh(PH_3)_2(CO)(\eta^2$-$CH_2{=}CHCH_3)$ complex was described at B3LYP level, whereas the bulky phosphines were described at UFF level. In their investigations, the catalyst has two monophosphines coordinated to the metal center. The authors performed a thorough investigation of all possible transition states that lead to the linear and branched products of the two families of feasible intermediate propene adducts, one with a bis-equatorial arrangement of the two PPh_3 ligands, and another with a mixed equatorial-axial arrangement.

In their studies the authors consider two possible scenarios for the formation of linear and branched aldehydes assuming an irreversible olefin insertion step. In the first scenario, the olefin insertion proceeds through a single intermediate, and the linear:branched product distribution is determined by the partitioning between the two forms (linear or branched) of the transition state (kinetic discrimination). In the second scenario, the linear and branched Rh-propyl insertion products arise from different intermediates (ground state discrimination); one of the propene adduct intermediates leads preferentially to the linear Rh-propyl insertion product, while another leads to the branched product.

Previous calculations showed that the propene prefers to be equatorial, whereas the H ligand prefers to be axial in the trigonal bipyramidal coordination complex. The authors considered six possible propene adducts as intermediates, two with both phosphines in equatorial positions, and four with one phosphine equatorial and the other axial. The two most stable intermediates account for ~90% of the Boltzmann population considering all of the possible intermediates. The most stable intermediate has phosphines in equatorial-axial positions, whereas the other one has the phosphines in equatorial-equatorial positions. Both intermediates have the methyl group of propene close to the hydride ligand.

To calculate the insertion transition states from each propene adduct, the authors considered the olefin rotation in clockwise and counterclockwise fashion for these two intermediates, as previously described by Carbó et al [115]. For the equatorial-axial adduct, the barrier to propene insertion leading to the linear insertion product was predicted to be 2.8 kcal/mol smaller than the barrier for the insertion reaction leading to the branched product. For the equatorial-equatorial adduct, the barrier for the insertion leading to the branched product was predicted to be 1.4 kcal/mol lower in energy than the barrier for the reaction leading to the linear product. Therefore, it appears that for this type of catalyst there are two separate propene insertion reaction channels, one generating almost exclusively the linear product, and the other producing primarily the branched product.

Decker and Cundari broadened their investigations of this reaction by studying the hydroformylation of ethylene by the $HRh(PR_3)_2(CO)$ complex for a variety of phosphines (PR_3; R=Me, ^{t}Bu, Ph, *meta*-$PhSO_3^-$, *para*-$PhSO_3^-$), in order to evaluate the steric influence of the PR_3 spectator ligand [118]. The ONIOM(B3LYP:UFF) approach was again used, including the ethylene and the $HRh(PH_3)_2(CO)$ complex in the QM part, and the rest of the system in the MM part, for each of the phosphine ligands. They studied the olefin insertion starting from the same intermediates as in the previous work. The difference in the activation barriers starting from both intermediates is quite small for all of the different phosphine ligands. Therefore, the authors suggest that both reaction channels will be operative for the ethylene insertion reaction in this catalyst system.

The two-layered ONIOM(B3LYP:HF) method describing the whole system at quantum mechanical level, without the support of a molecular mechanics method, was employed by Landis and Uddin to explore the hydroformylation of 1-alkenes by a Rh-xantphos complex [119]. The authors state that their results dispel the assumption that only phosphine diequatorial isomers of the critical intermediates of the reactions contribute to product formation, because the equatorial-axial isomers are essentially isoenergetic.

These studies on the hydroformylation of olefins by rhodium catalysts, initially performed at QM level and later extended using QM/MM methods, represent substantial progress in the comprehension of the regio- and stereoselectivity of this reaction. At any rate, subsequent experimental and computational work is needed to achieve a full understanding of the overall mechanism.

6 Other Transition Metal Catalyzed Reactions

The previous sections were dedicated to the transition metal catalytic reactions that have been widely studied using QM/MM methods. In this section, other transition metal catalyzed reactions that have also had QM/MM methods applied to them are briefly discussed.

The enantioselective addition of dialkylzinc to benzaldehyde catalysed by several fenchone derivatives has been studied by Goldfuss et al using the ONIOM(RHF:UFF) method [120]. The authors studied these reactions experimentally, and used computational analysis to rationalize the results. Examinations of the "anti" and "syn" transition structures for the addition show how the interactions between the fenchone-based ligands and the alkyl groups attached to the zinc affect the relative stabilities of the transition states. The results are in good agreement with experiment, and show how computational studies can help with ligand design. In a related work, the enantioselective addition of diethylzinc to benzaldehyde mediated by (*R*)-2-piperidino-1,1,2-triphenylethanol was studied via the IMOMM(HF:MM3) method by Vázquez et al [121]. In this study, the authors consider different QM/MM partitions, as well as the complete ethyl substituents. They found that the different partitions produced similar results, validating the partition with the smaller QM region. Moreover, the nontransferred ethyl groups bonded to zinc are found to be crucial to stereochemical discrimination during the reaction.

Cavallo and Solà [122] carried out a full QM and QM/MM study of steric and electronic effects in rhodium-catalyzed carbonylation reactions. The investigations concerned the two main steps of the reaction: the oxidative addition of the alkyl halide (the rate determining step), and the CO insertion step. The reaction was investigated for several $Rh(CO)I(L)_2$ complexes, where L represents various phosphine or mixed bidentate ligands. The systems considered adopt a square-planar geometry prior to the CH_3I oxidative addition. Looking at the octahedral complexes formed after the oxidative addition, the results indicate that the energy gain due to the addition is reduced considerably by the steric pressure of the substituents on the ligand.

Similar trends are obtained using selected pure QM calculations on the real system, a result indicating that electronic effects of substituents are not the main influence on the energetics of the oxidative addition. As far as the CO insertion step is concerned, results show that the energy barrier is lowered in the presence of bulky ligands. This effect is related to a relief in the steric pressure on the complex as the system moves from a six-coordinated octahedral geometry toward a five-coordinated square-pyramidal geometry. In a related computational work, Bo and coworkers [123] studied the electronic and steric effects in the migratory CO insertion step for $RhMe(CO)I_2(L\text{-}L)$ catalysed methanol carbonylation, where L-L are phosphine (dppe) and mixed sulfide phosphine (dppms) ligands. The difference in the reaction rates for both ligands was suggested to be of electronic origin, from the different properties of the sulfide phosphine (π-donor) and diphosphine (π-acceptor) ligands. Steric effects, in turn, were found to modulate the barrier, increasing it more for dppe than for dppms. The agreement between the calculated and experimental energy barriers for both ligands was excellent.

The catalytic epoxidation of prochiral olefins with Mn(salen) complexes (H_2salen=bien(salicylidene)ethylene-diamine), also known as the Jacobsen-Katsuki reaction, was investigated by Jacobsen and Cavallo [124] using a QM/MM method. The authors concentrate on the oxidation of *cis*-β-methylstyrene, suggesting that the enantioselectivity of the reaction is related to three specific features: (i) the chiral diimine bridge of the salen ligand induces a chiral pocket; (ii) the bulky groups of the salen ligand cause a preferential approach of the aromatic ring of the olefin; (iii) the substituents of the olefin confer regioselectivity and therefore enantioselectivity.

Tobisch and Ziegler [125] performed a theoretical investigation of the influence of the ligand L on the regulation of product selectivity for the $[Ni^0L]$-catalyzed cyclodimerization of 1,3-butadiene. The catalytic cycle was analyzed for several ligands: L=PMe_3, PPh_3, $P(^iPr)_3$ and $P(OPh)_3$. The role of electronic and steric effects was elucidated for all crucial elementary steps of the entire catalytic cycle. The electronic and steric effects for the first steps of the catalytic cycle, the oxidative coupling of two butadienes, analyzed using a QM/MM method, are shown to be small. Consequently, the activation barriers are very similar for all catalysts, indicating that this step is nearly independent of the ligand's nature. Cyclodimer products are formed along competing routes for reductive elimination, which is predicted to be the rate-determining step. The barriers for these steps are shown to rise upon increasing the ligand's σ-donor strength, but not uniformly. These results give a detailed and fully consistent explanation for the catalytic reaction cycle.

Adlhart and Chen [126] reported a QM/MM study of the olefin-metathesis reaction catalyzed by $[L_2(Cl)_2Ru{=}CHPh]$ complexes; L are phosphane groups in the case of the "first-generation" catalyst, whereas for the "second-generation" catalysts an N-heterocyclic carbene ligand replaces one of the phosphanes. A study of first- and second-generation metathesis catalysts is presented, and they find that the major topological differences between the potential surfaces for both generations of catalysts can be traced to the different symmetries of the phosphane and N-heterocyclic carbene ligands. The conclusions of their study allowed them to propose a new ligand that combines the advantages of both generations of catalysts.

Enantioselective hydrosilylation by Pd species was studied by Woo and coworkers [127] through ab initio molecular dynamics and QM/MM simulations. The evaluation of a series of combined QM/MM model systems allowed the steric and electronic influences of specific ligands to be determined. Therefore, the nature of the geometric distortions produced by these ligands were pinpointed. Woo and coworkers found that the phosphine phenyl substituents contribute to the Pd-P lengthening on both steric and electronic levels. With an examination of the complete catalytic cycle, the authors were able to explain the high regioselectivity observed, and to rule out some of the reaction mechanisms suggested by experimental results.

Aside from full catalytic cycles, or reaction steps within them, as discussed above, structural issues about organometallic complexes with catalytic properties have also been analyzed with QM/MM methods. For instance, Helmchen and coworkers carried out a conformational analysis of two (π-1,3-dimethylallyl)(phosphinooxazoline)Pd complexes [128], and Magistrato et al investigated the role of π-π stacking interactions in square planar palladium complexes [129].

7 Concluding Remarks

Results presented in this contribution prove that the application of DFT/MM computational methods to homogeneous catalysis is a young yet very active field of research. A detailed knowledge of reaction mechanisms is highly desirable in homogeneous catalysis, because it facilitates the job of catalyst optimization and design, especially when the practical focus is on high selectivity associated with bulky ligands. Unfortunately, the reactions are usually complicated, with little detailed associated experimental data. Their full understanding often requires the consideration of a variety of possibilities, as well as a critical evaluation of experimental data. The interaction between experimental and computational work appears to be a promising way forward in this area.

DFT/MM methods currently seem to offer the best balance between computational cost and accuracy for the computational study of many of the systems of experimental interest to homogeneous catalysis. This does not mean that the theoretical work is always easy. In fact, the results summarized in this contribution show a varying degree of progress in the understanding of the different reactions, usually associated with the amount of effort invested. There is already a good general mechanistic understanding of olefin polymerization, the reaction that has been investigated in the most detail, and current work, both computational and experimental, is focusing on the tuning of the catalysts to give the specific required products and substrates. In other processes, like those presented in the last section, the knowledge is still quite preliminary, and more work is required before the mechanistic information can be used for a practical purpose.

The application of DFT/MM methods to homogeneous catalysis complements the use of pure DFT methods. On the one hand, a detailed knowledge of the main features of the reaction mechanism is mandatory before a DFT/MM study, and this is often best accomplished with a pure DFT study on a model system. On the other hand, pure QM calculations on selected systems remain the benchmark for the reliability of DFT/MM results.

In fact, one of the strong points of these DFT/MM methods is that they profit directly from methodological progress in either DFT or MM methods, as well as from progress in the DFT/MM combination itself. This progress, foreseeable in the near future, as well as increased computing power, will surely contribute to a rapid increase in the use of the DFT/MM approach for practical problems in homogeneous catalysis.

References

1. Cornils B, Herrman WA (2002) (eds) Applied homogeneous catalysis with organometallic compounds, 2nd edn. Wiley-VCH, Weinheim
2. Jacobsen EN, Pfaltz A, Yamamoto H (1999) (eds) Comprehensive asymmetric catalysis. Springer, Berlin Heidelberg New York
3. Koch W, Holthausen MC (2000) A chemist's guide to density functional theory. Wiley-VCH, Weinheim
4. Comba P (1999) Coord Chem Rev 182:343
5. Maseras F, Lledós A (2002) (eds) Computational modeling of homogeneous catalysis. Kluwer, Dordrecht
6. Torrent M, Solà M, Frenking G (2000) Chem Rev 100:439
7. Ziegler T (2002) J Chem Soc Dalton Trans 642
8. Warshel A, Levitt, M (1976) J Mol Biol 103:227
9. Monard G, Merz K (1999) Acc Chem Res 32:904
10. Truhlar DG, Gao J, Alhambra C, Garcia-Viloca M, Corchado J, Sánchez ML, Villà J (2002) Acc Chem Res 35:341
11. Maseras F, Morokuma K (1995) J Comput Chem 16:1170
12. Svensson M, Humbel S, Froese RDJ, Matsubara T, Sieber S, Morokuma K (1996) J Phys Chem 114:10024
13. Dapprich S, Komáromi I, Byun KS, Morokuma K, Frisch MJ (1999) J Mol Struct–THEOCHEM 461–462:1
14. Sauer J, Sierka M (2000) J Comput Chem 21:1470
15. Maseras F (2000) Chem Commun 1821
16. Barea G, Maseras F, Jean Y, Lledós A (1996) Inorg Chem 35:6401

17. Jaffart J, Etienne M, Maseras F, McGrady JE, Eisenstein O (2001) J Am Chem Soc 123:6000
18. Lam BMT, Halfen JA, Young VG, Hagadorn JR, Holland PL, Lledós A, Cucurull-Sánchez L, Novoa, JJ, Alvarez S, Tolman WB (2000) Inorg Chem 39:4059
19. Gusev DG, Lough AJ (2002) Organometallics 21:5091
20. Maseras F (1999) Top Organomet Chem 4:165
21. Maseras F (2001) In: Cundari TR (ed) Computational organometallic chemistry. Marcel Dekker, New York
22. Balcells D, Drudis-Solé G, Besora M, Dölker N, Ujaque G, Maseras F, Lledós A (2003) Faraday Discuss 124:429
23. Kaminsky W, Arndt-Rosenau M (2002) In: Cornils B, Herrmann WA (eds) Applied homogeneous catalysis with organometallic compounds, 2nd edn. Wiley-VCH, Weinheim, p 213
24. Ziegler K, Holzkamp E, Breil H, Martin H (1955) Angew Chem 67:541
25. Natta G (1956) Angew Chem 68:393
26. Andresen A, Cordes HG, Herwig J, Kaminsky W, Merck A, Mottweiler R, Pein J, Sinn H, Vollmer HJ (1976) Angew Chem Int Ed Engl 15:630
27. Sinn H, Kaminsky W (1980) Adv Organomet Chem 18:99
28. Resconi L, Cavallo L, Fait A, Piamontesi F (2000) Chem Rev 100:1253
29. Ittel SD, Johnson LK, Brookhart M (2000) Chem Rev 100:1169
30. Gibson VC, Spitzmesser SK (2003) Chem Rev 103:283
31. Cossee P (1964) J Catal 3:80
32. Rappe AK, Skiff WM, Casewit CJ (2000) Chem Rev 100:1435
33. Cavallo L (2002) In: Maseras F, Lledós A (eds) Computational modeling of homogeneous catalysis. Kluwer, Dordrecht, p 23
34. Michalak A, Ziegler T (2002) In: Maseras F, Lledós A (eds) Computational modeling of homogeneous catalysis. Kluwer, Dordrecht, p 57
35. Lauher JW, Hoffmann R (1976) J Am Chem Soc 98:1729
36. Deng L, Woo TK, Cavallo L, Margl PM, Ziegler T (1997) J Am Chem Soc 119:6177
37. Froese RDJ, Musaev DG, Morokuma K (1998) J Am Chem Soc 120:1581
38. Musaev DG, Froese RDJ, Morokuma K (1998) Organometallics 17:1850
39. Johnson LK, Killian CM, Brookhart M (1995) J Am Chem Soc 117:6414
40. Woo TK, Margl PM, Blöchl PE, Ziegler T (1997) J Phys Chem B 101:7877
41. Woo TK, Blöchl PE, Ziegler T (2000) J Phys Chem A 104:121
42. Deng L, Margl P, Ziegler T (1999) J Am Chem Soc 121:6479
43. Margl P, Deng L, Ziegler T (1999) Organometallics 18:5701
44. Khoroshun DV, Musaev DG, Vreven T, Morokuma K (2001) Organometallics 20:2007
45. Ramos J, Cruz V, Muńoz-Escalona A, Martínez-Salazar J (2002) Polymer 43:3635
46. Deng L, Ziegler T, Woo TK, Margl P, Fang L (1998) Organometallics 17:3240
47. Chan MSW, Deng L, Ziegler T (2000) Organometallics 19:2741
48. Vyboishchikov SF, Musaev, DG, Froese RDJ, Morokuma K (2001) Organometallics 20:309
49. Lanza G, Fragala IL, Marks TJ (1998) J Am Chem Soc 120:8257
50. Xu Z, Vanka K, Firman T, Michalak A, Zurek E, Zhu C, Ziegler T (2002) 21:2444
51. Cavallo L, Guerra G, Vacatello M, Corradini P (1991) Macromolecules 24:1784
52. Guerra G, Cavallo L, Moscardi G, Vacatello M, Corradini P (1994) J Am Chem Soc 116:2988
53. Hart JR, Rappe AK (1993) J Am Chem Soc 115:6159
54. Kawamura-Kuribayashi H, Koga N, Morokuma K (1992) J Am Chem Soc 114:8687
55. Moscardi G, Resconi L, Cavallo L (2001) Organometallics 20:1918

56. Guerra G, Longo P, Corradini P, Cavallo L (1999) J Am Chem Soc 121:8651
57. Longo P, Grisi F, Guerra G, Cavallo L (2000) Macromolecules 33:4647
58. Longo P, Pragliola S, Milano G, Guerra G (2003) J Am Chem Soc 125:4799
59. Busico V, Castelli VVA, Aprea P, Cipullo R, Segre A, Talarico G, Vacatello M (2003) J Am Chem Soc 125:5451
60. Milano G, Cavallo L, Guerra G (2002) J Am Chem Soc 124:13368
61. Talarico G, Busico V, Cavallo L (2003) J Am Chem Soc 125:7172
62. Osborn JA, Jardine FH, Young JF, Wilkinson G (1966) J Chem Soc A 1711
63. Knowles WS, Sabacky MJ (1968) J Chem Soc Chem Commun 1445
64. Horner L, Siegel H, Büthe H (1968) Angew Chem Int Ed Engl 7:942
65. Dang TP, Kagan HB (1972) J Am Chem Soc 94:6429
66. Miyashita A, Yasuda A, Takaya H, Toriumi K, Ito T, Souchi T, Noyori R (1980) J Am Chem Soc 102:7932
67. Burk MJ (1991) J Am Chem Soc 113:8518
68. Knowles WS (1983) Acc Chem Res 16:106
69. Chan ASC, Halpern J (1980) J Am Chem Soc 102:838
70. Brown JM, Parker D (1982) Organometallics 1:950
71. Armstrong SK, Brown JM, Burk MJ (1993) Tetrahedron Lett 34:879
72. Gridnev ID, Yasutake M, Higashi N, Imamoto T (2001) J Am Chem Soc 123:5268
73. Landis CR, Hilfenhaus P, Feldgus S (1999) J Am Chem Soc 121:8741
74. Feldgus S, Landis C (2002) In: Maseras F, Lledós A (eds) Computational modeling of homogeneous catalysis. Kluwer, Dordrecht, p 107
75. Landis CR, Hilfenhaus P, Feldgus S (1999) J Am Chem Soc 121:8741
76. Esteruelas MA, Oro LA (1998) Chem Rev 98:577
77. Daniel D, Koga N, Han J, Fu XY, Morokuma K (1988) J Am Chem Soc 110:3773
78. Gridnev ID, Higashi N, Asakura K, Imamoto T (2001) Adv Synth Catal 343:118
79. Feldgus S, Landis CR (2000) J Am Chem Soc 122:12714
80. Landis CR, Feldgus S (2000) Angew Chem Int Ed Engl 39:2863
81. Feldgus S, Landis CR (2001) Organometallics 20:2374
82. Sharpless KB (2002) Angew Chem Int Ed Engl 41:2024
83. Kolb HC, VanNieuwenhze MS, Sharpless KB (1994) Chem Rev 94:2483
84. (a) Corey EJ, Noe, MC, Sarshar, S (1993) J Am Chem Soc 115:3828; (b) Corey EJ, Noe MC (1996) J Am Chem Soc 118:11038
85. (a) Göble T, Sharpless KB (1993) Angew Chem Int Ed Engl 32:1329; (b) Norrby PO, Becker H, Sharpless KB (1996) J Am Chem Soc 118:35
86. (a) Dapprich S, Ujaque G, Maseras F, Lledós A, Musaev DG, Morokuma K (1996) J Am Chem Soc 118:11660; (b) Pidun U, Boehme C, Frenking G (1996) Angew Chem Int Ed 35:2817; (c) Torrent M, Deng L, Duran M, Solà M, Ziegler T (1997) Organometallics 16:13; (d) Del Monte AJ, Haller J, Houk KN, Sharpless KB, Singleton DA, Strassner T, Thomas AA (1997) J Am Chem Soc 119:9907
87. Deubel DV, Frenking G (2003) Acc Chem Res 36:645
88. Deubel DV, Schlecht S, Frenking G (2001) J Am Chem Soc 123:10085
89. Gable KP, Zhuravlev FA (2002) J Am Chem Soc 124:3970
90. Ujaque G, Maseras F, Lledós A (2003) Eur J Org Chem 5:833
91. Norrby P-O, Rasmussen T, Haller J, Strassner T, Houk KN (1999) J Am Chem Soc 121:10186
92. (a) Norrby P-O, Liljefors T (1998) J Comput Chem 19:1146 (b) Norrby P-O (1999) In: Truhlar DG, Morokuma K (eds) Transition state modeling for catalysis, ACS Symposium Series, No 721. ACS, Washington DC, p 163

93. (a) Moitessier NN, Maigret B, Chretien F, Chapleur Y (2000) Eur J Org Chem 995; (b) Moitessier N, Henry C, Len C, Chapleur Y (2002) J Org Chem 67:7275
94. (a) Ujaque G, Maseras F, Lledós A (1996) Theor Chim Acta 94:67; (b) Ujaque G Maseras F, Lledós A (1997) J Org Chem 62:7892
95. Ujaque G, Maseras F, Lledós A (1999) J Am Chem Soc 121:1317
96. Kolb HC, Andersson PG, Sharpless KB (1994) J Am Chem Soc 116:1278
97. Drudis-Solé G, Ujaque G, Maseras F, Lledós A (2004) Chem Eur J (submitted)
98. Corey EJ, Noe MC (1993) J Am Chem Soc 115:12579
99. Leach AF (2001) Molecular modelling, principles and applications, 2nd edn. Pearson, Harlow, UK
100. Nishio M, Hirota M, Umezawa Y (1999) The C-H/π interaction: evidence, nature and consequences. Wiley-VCH, New York
101. Pratt LR, Pohorille A (2002) Chem Rev 102:2671
102. van Leeuwen PWNM, Claver C (2000) (eds) Rhodium catalyzed hydroformylation. Kluwer, Dordrecht
103. Herrmann WA, Cornils B (1997) Angew Chem Int Edit 36:1047
104. Evans D, Osborn JA, Wilkinson G (1968) J Chem Soc (A) 3133
105. (a) Grima JP, Choplin F, Kaufman G, (1977) J Organomet Chem 129:221; (b) Antolovic D, Davidson ER (1987) J Am Chem Soc 109:5828; (c) Veillard A, Daniel C, Rohmer M-M (1990) J Phys Chem 94:5556
106. Versluis L, Ziegler T, Fan L (1990) Inorg Chem 29:4530
107. Pidun U, Frnking G (1998) Chem Eur J 4:522
108. Matsubara T, Koga N, Ding Y, Musaev DG, Morokuma K (1997) Organometallics 16:1065
109. Decker SA, Cundari TR (2001) Organometallics 20:2827
110. Rocha WR, de Almeida WB (2000) Int J Quantum Chem 78:42
111. Castonguay LA, Rappe AK, Casewit CJ (1991) J Am Chem Soc 113:7177
112. (a) Gleich D, Schmid R, Herrmann, WA (1998) Organometallics 17:2141; (b) Gleich D, Schmid R, Herrmann, WA (1998) Organometallics 17:4828
113. Casey CP, Petrovich LM (1995) J Am Chem Soc 117:6007
114. Gleich D, Herrmann WA (1999) Organometallics 18:4354
115. Carbó JJ, Maseras F, Bo C, van Leeuwen PWNM (2001) J Am Chem Soc 123:7630
116. Carbó JJ, Maseras F, Bo C (2002) In: Maseras F, Lledós A (eds) Computational modeling of homogeneous catalysis. Kluwer, Dordrecht, p 161
117. Decker SA, Cundari TR (2001) J Organomet Chem 635:132
118. Decker SA, Cundari TR (2002) New J Chem 26:129
119. Landis RC, Uddin J (2002) J Chem Soc Dalton Trans 729
120. (a) Goldfuss B, Steigelmann M, Khan SI, Houk KN (2000) J Org Chem 65:77; (b) Goldfuss B, Steigelmann M, Rominger F (2000) Eur J Org Chem 1785
121. Vázquez J, Pericàs MA, Maseras F, Lledós A (2000) J Org Chem 65:7303
122. Cavallo L, Solà M (2001) J Am Chem Soc 123:12294
123. Daura-Oller E, Poblet JM, Bo C (2003) Dalton T 92
124. Jacobsen H, Cavallo L (2001) Chem Eur J 7:800
125. Tobisch S, Ziegler T (2002) J Am Chem Soc 124:13290
126. Adlhart C, Chen P (2002) Angew Chem Int Edit 41:4484
127. Magistrato A, Togni A, Rothlisberger U, Woo TK (2002) In: Maseras F, Lledós A (eds) Computational modeling of homogeneous catalysis. Kluwer, Dordrecht, p 213
128. Vázquez J, Goldfuss B, Helmchen G (2002) J Organomet Chem 641:67
129. Magistrato A, Pregosin PS, Albinati A, Rothlisberger U (2001) Organometallics 20:4178

Structure and Bonding (2004) 112:151–183
DOI 10.1007/b97939

DFT Computation of Relative Spin-State Energetics of Transition Metal Compounds

Jeremy N. Harvey

School of Chemistry, University of Bristol, Cantock's Close, Bristol, BS8 1TS, UK
jeremy.harvey@bris.ac.uk

Abstract DFT is often used to predict the energetics of transition metal compounds. In particular, energy differences between states of different spin are of great interest. This review discusses the accuracy of such computations for spin-state splittings of monometallic complexes where the differences in spin-pairing occur mostly or exclusively on the metal itself. Large differences are found for computed splittings using different functionals, in particular between pure and hybrid functionals. An optimum exact exchange admixture of ~15% seems to yield accurate results in many cases, but experimental and especially theoretical uncertainties suggest that this should not be relied upon, and the development of new functionals is seen to be highly desirable.

Keywords DFT · Spin-state · Transition Metals · Exact exchange · Self-interaction

List of Abbreviations

TM	Transition Metal
DFT	Density functional theory
HF	Hartree-Fock

1 Introduction

The problem of predicting the structure of and the nature of bonding in stable organic or main-group inorganic compounds is comparatively straightforward in the sense that the qualitative features of the electronic wavefunction of the ground state of the species are usually known. Most of these molecules have closed-shell configurations, with all of their electrons paired up to give a singlet spin state. Of course, even in this area of chemistry there are exceptions to this rule. For example, many "free radicals", with one unpaired electron, are known, and although many of these are intermediates with a low lifetime, some can be isolated in pure form. There are also several well-known types of reactive intermediate with more than one unpaired electron. For example, divalent carbon compounds, or carbenes, have two electrons on the carbon atom that do not participate directly in bonding, and which can occupy two orbitals. As a result, carbenes often have several close-lying electronic states, the most important of which are the open-shell triplet, and the closed-shell singlet. In fact, depending on the substituents on the carbon, either of these can be the ground state. Such systems are however exceptions within organic and main group chemistry. For most compounds in these fields, the nature of the electronic ground state is fairly predictable. This is not to deny that there are many other difficulties associated with theoretical studies of their structure and bonding!

Transition metal (TM) chemistry stands in contrast to this. Many compounds involve metal centres with partially filled *d* shells, and/or with one or several unpaired electrons. Therefore, it is not always straightforward to predict the orbital occupation pattern of a given stable compound. For intermediates on a reactive pathway, this is an even greater problem. This is also true for organometallic chemistry, despite the fact that many compounds obey the 18-electron rule and have closed-shell singlet ground states. Thus, there are many 16- or even 14-electron intermediates, odd-electron species [1], and polymetallic clusters and complexes for which the spin state is not readily predicted.

Computational and theoretical chemistry has a very important role to play in helping to predict and rationalize the nature of the electronic ground state of TM compounds. Being able to do so is critical in many respects, if one wishes to predict the structure, properties, and reactivity of such compounds. First of all, the predicted structure and properties (for example the spectroscopic features) will of course be very different for different spin states. Of more interest to our research group is the notion that *reactivity* often crucially depends on the preferred spin state of reactants, products and intermediates. Thus, we have shown elsewhere (for reviews, see [2–4]) that many reactions of TM compounds involve multiple electronic states, of different spin. In these cases, reactivity is strongly influenced by the energy

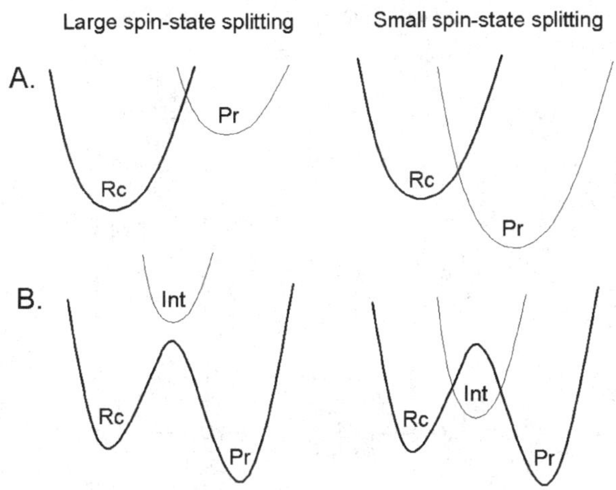

Fig. 1 Two cases show how changes in predicted spin-state splittings can affect the predicted quantitative and qualitative features of a reactive system. In A, reactants (Rc) and products (Pr) belong to different spin states, whereas in B, they have the same spin state, but a potential intermediate (Int) has a different spin

splitting between the different spin states. Two generic examples of this are shown schematically in Fig. 1. In example 1, shown in the upper part of the Figure, the ground state of the reactants (Rc) belongs to one spin state, shown as a bold potential energy curve, and the ground state of the products (Pr) lies on another spin-state surface, shown as a lighter line. In Case A, the reaction is endothermic and involves a significant barrier, whereas in Case B, in which the second spin state is (relatively) much lower in energy, the reaction is exothermic and has a low barrier. Example 2 involves an intermediate (Int) which differs in spin state from reactants and products. In Case A, this spin state lies at such high energy that it will not be formed, and the reaction will proceed adiabatically over a high barrier. In Case B, the spin-state is relatively much more stable, and it now becomes more favourable, from the energetic point of view, to have a double spin-state change during the reaction. Clearly, in order to be able to predict the qualitative features of such reactions, it is necessary to be able to predict relative spin-state energetics accurately!

For all of these reasons, transition metal chemists have developed many semi-empirical and ab initio, qualitative and semi-quantitative procedures to try to predict the ordering of electronic states. The purpose of the present review is to survey the success of one of the most sophisticated of these methods, density functional theory, in calculating the relative energetics of spin states.

Compounds can have close-lying electronic states of different spin for several reasons. For convenience, we have divided these into three partially

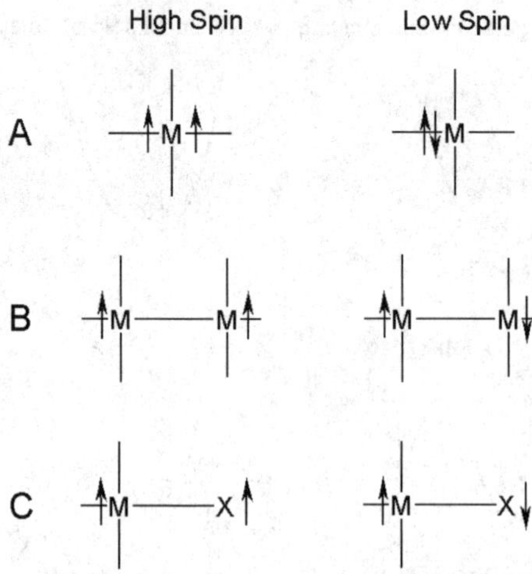

Fig. 2 Different scenarios for obtaining multiple spin-states in a transition metal compound. M is a metal atom with generic ligands, X is a non-metallic group bearing an unpaired electron

overlapping categories, as shown in Fig. 2. In case A, electrons on the metal (usually within the *d* shell) couple up in different ways to give either a high or a low (or an intermediate) spin overall. This very common case is very familiar, for example in crystal field theory. Many examples will be discussed below. In case B, two (or more) metal centres are present in the molecule (or material), and both of them have one or more unpaired electrons contributing to a "local" spin angular momentum. These momenta then couple, either through space or through the bonding orbitals of the "bridge" connecting the metals, to provide the overall spin. This case is well-known from solid state chemistry of substances such as metal oxides, with parallel and antiparallel arrangements of the metal ion spin components leading to either ferromagnetic or antiferromagnetic solids. Many bi- and polymetallic molecular complexes are also known to display this kind of coupling, and there is considerable experience in using DFT methods to predict the spin-state splitting in such compounds (see reviews: [5–7]).

Likewise, in the final case, C, spin angular momentum on a metallic centre can be more or less closely situated with respect to an unpaired electron (or electrons) situated on a non-metallic centre. One example of this situation concerns the Compound I intermediate of oxidative cytochrome P450 enzymes. This compound involves two unpaired electrons on a formally Fe(IV)-O iron oxo core, high-spin coupled to form a local $S=1$ spin contribution, with a third unpaired electron residing in the π orbitals of the porphyrin ligand [8]. Coupling between the two spin components can lead to both doublet and quartet states.

The bulk of this review will focus on situations belonging to category A, without addressing the very different issues involved with cases B and C in which the inter-spin coupling is much weaker. Nevertheless, some systems occur at the borderline between A and either B or C, and for these, the present review will be of some relevance.

2
Wavefunction and DFT Methods

2.1
Wavefunction Methods

Before reviewing their specific behaviour for the problem at hand, it is appropriate to briefly remind the reader of the mathematical form of Hartree-Fock (HF) and of Density Functional (DFT) theories (for more detail on the material contained in this section, the reader is referred to one of the many excellent books on electronic structure theory, such as [9]). HF expresses the electronic wavefunction as a Slater determinant – an antisymmetrized product of one-electronic functions, or molecular orbitals. For a given system defined by the nature and position of the (fixed) atomic nuclei and by the number of electrons n, the electronic energy, E_{elec} (to which the nuclear-nuclear repulsion energy must be added to give the total energy, which defines the potential energy surface), is given as:

$$E_{\mathrm{elec}} = \sum_{i=1}^{n} (t_{\mathrm{i}} + V_{\mathrm{Ni}}) + \frac{1}{2}\sum_{\mathrm{i}=1}^{n}\sum_{j=1}^{n} (J_{\mathrm{ij}} - K_{\mathrm{ij}}) \tag{1}$$

Here t_{i} is the kinetic energy of an electron in molecular orbital i; V_{Ni} is the interaction energy of the same electron with the Coulombic potential induced by the positively charged nuclei; the "Coulomb" integral J_{ij} is the classical interaction energy between the charge density corresponding to an electron in orbital i and that corresponding to another electron in orbital j; and the "Exchange" integral K_{ij} serves to correct the Coulomb integral for its neglect of the wavefunction antisymmetrization. The mathematical form of K_{ij} means that it is equal to zero for electrons of opposite spin. It is of some importance for the discussion of DFT, below, to note that the exchange integrals in Eq. 1 have a second important function. Thus, the double summation includes the terms where $i=j$ (in other words the energy includes the repulsion energy of each electron with itself!). Fortunately, J_{ii} is exactly equal to K_{ii}, so that this "self-interaction" contribution cancels out exactly. This may appear to be a mere technical detail concerned with the way in which the summation of the J_{ij} and K_{ij} contributions is carried out, but it turns out that the precise cancellation is required for HF calculations to be carried out

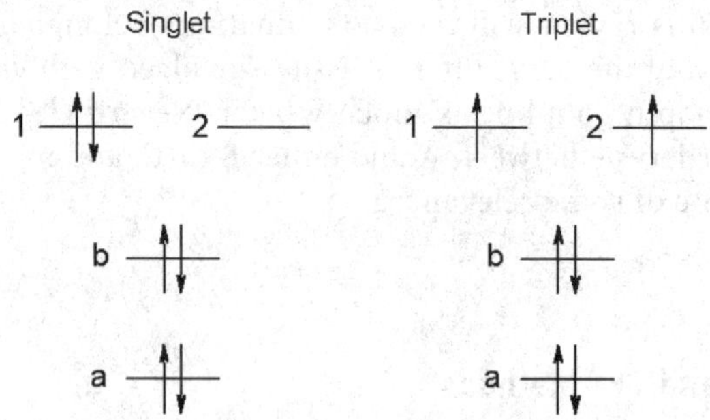

Fig. 3 Schematic molecular orbital diagram for singlet and triplet spin states of a metal complex, showing doubly-occupied (ligand-based, for example) molecular orbitals a and b, and a degenerate pair of metal orbitals, 1 and 2

effectively. It would not be convenient to need to exclude all the terms where $i=j$.

From this discussion of HF theory, it is possible to discuss the expected calculated energetics for different spin states of metal complexes. For case A, above, the electronic configuration of the two spin states can be shown schematically as in Fig. 3. Making a few simplifying assumptions, the HF difference in energy between the singlet and triplet states can be given as:

$$\Delta E_{\mathrm{HF}}^{1,3} = (J_{11}) - (J_{12} - K_{12}) \tag{2}$$

This assumes that the form of the molecular orbitals is the same in both states, that the one-electron parts of the energy expressions are the same for both states, and that all of the Coulomb and Exchange integrals involving one of the two electrons in orbitals 1 and 2 with one of the other electrons are identical to those involving the other of the two electrons (that is, that $J_{1a}=J_{2a}$, and so on). If both orbitals are mainly *d* atomic orbitals localized on the metal atom, then J_{11} and J_{12} will be of the same order and the energy difference will be given by K_{12}, the exchange integral or "interaction" between the two electrons in the triplet state. Since this energy contribution is positive (and fairly large in chemical terms), the triplet state will be significantly lower than the singlet. The cartoon in Fig. 3 shows both *d* orbitals as isoenergetic. In cases where one orbital lies lower than the other (for example due to better interactions with the nuclei on a ligand or to less electron-electron repulsion with ligands), the triplet will still usually come out as lower in HF calculations, unless the orbital splitting is very large.

Although some of the assumptions made are certainly not correct in detail, this qualitative prediction of how HF behaves for the spin splitting in metal complexes of type A is in good agreement with the results of HF calculations (see below). In almost all cases, the high-spin state is far too stable with respect to the low-spin one when the splitting is compared with experimental data. This is because HF neglects an important physical effect: electron correlation. Because of Coulombic repulsion, the probability of finding one electron at a given point in space depends on whether or not another electron is already in the vicinity. HF theory neglects part of this effect. If the wavefunction were written as a simple product of molecular orbitals (Hartree product), the neglect would in fact be total, with each electron moving completely independently of the others. The Slater determinant, however, includes some "correlation" in that the probability of finding two electrons of *same spin* near each other is decreased – antisymmetrization introduces "Fermi holes" around each electron in which the probability of finding another electron of the same spin is lower than would be predicted by the Hartree product. However, HF does not describe the similar effect that occurs between electrons irrespective of their spin, the "Coulomb hole". To do so, it is necessary to expand the electronic wavefunction as a linear combination of multiple Slater determinants. The larger the expansion, the closer to the exact result (barring effects such as relativity which are not included in the simplest treatments).

It is well-known that wavefunction-based methods, whilst in principle, of course, capable of reaching the exact result in this way, struggle to reproduce experimental spin-state splittings for transition metal compounds (for examples, see below). This is because of the large size and therefore huge computational cost of the expansion needed to achieve convergence on the energy difference. Also, simple methods to include electron correlation, such as MP2 theory, which are often qualitatively successful for main group compounds, sometimes fail dramatically for TM systems (for reviews of ab initio calculations on TM compounds, see [10] for example). Even for very small TM compounds, it can be rather difficult to achieve quantitative accuracy using ab initio methods.

2.2 Density Functional Theory

Compared to wavefunction methods, DFT follows an apparently completely different approach in order to solve the same problem – the determination of the energy of a collection of fixed nuclei and moving electrons (for general introductions to DFT, see [9], as well as more specialized sources such as [11]). Instead of the many-dimensional wavefunction, the central object in DFT is a comparatively simple function of three variables, the electronic density ρ within the molecular space. DFT is built on two fundamental theorems proposed by Hohenberg and Kohn. The first states that there exists in principle a *functional* $E[\rho]$ from which the exact energy of the system can be computed upon inputting the exact density function ρ. The second theorem states that this functional is variational: the energy computed from a given density function is lower the closer the input density is to the correct one.

The problem in DFT is to find the correct form for the functional, and, using that, to solve for the density.

The exact functional of DFT is unknown, and, even if it was known, there is no guarantee that it would be significantly more computationally tractable to evaluate energies and densities from it than it is to obtain them from exact solutions of the Schrödinger equation within the wavefunction approach. However, many *approximate* functionals are known which give excellent results for a range of chemical problems, and it is with those that we are concerned. All of these functionals adopt the same basic energy expression:

$$E_{\text{elec}}(\rho) = T(\rho) + V_{\text{Ne}}(\rho) + J_{\text{ee}}(\rho) + E_{\text{xc}}(\rho) \quad (3)$$

The four functional terms to the right of this equation correspond respectively to the kinetic energy of the electrons in the density distribution, the Coulombic interaction of the electron distribution with the nuclei (or with other external fields), the classical Coulombic interactions of the electrons amongst themselves, and the "exchange-correlation" functional which corrects the former three terms for their neglect of various effects. For practical reasons (see below), it is convenient to treat the density as arising from a sum of contributions from a set of doubly- (or singly-) occupied molecular (Kohn-Sham) orbitals. Applying the variational condition to E_{elec} therefore involves finding the set of orbitals that minimise the energy; doing so leads to the Kohn-Sham equation which in outward form is very similar to the Fock equation of HF theory, and also needs to be solved self-consistently.

Whilst Eq. 3 is in principle *exact*, practical implementations necessarily involve approximations. To get a better understanding of the impact of these approximations on spin-state energetics, it is useful to discuss each term in some more detail.

The first term, the kinetic energy, is difficult to calculate directly from the density, and it is for this reason that the "molecular orbitals" mentioned above are introduced: a very good approximation to the kinetic energy corresponding to the density can be calculated from the orbitals as it would be in HF theory. This approach does *not* however yield the *exact* kinetic energy because it assumes that the electrons in each orbital do not interact with electrons in other orbitals. The exact exchange-correlation functional must therefore contain a corrective term to incorporate the effect of electronic interactions on their kinetic energy. In practice, such a term is not explicitly included in common functionals.

The second term, describing the interaction of the electron density with the positive nuclei (or other external fields), is straightforward.

The third term, describing the interelectronic Coulomb repulsion energy, is given in density functional theory as:

$$J_{ee}(\rho) = \frac{1}{2}\iint \frac{\rho(r)\rho(r')}{|r-r'|}drdr' \tag{4}$$

This is in fact mathematically identical to the sum of the J_{ij} Coulomb integrals discussed above in the context of HF theory. As also mentioned there, this is *not* the true electron-electron repulsion energy, for several reasons. First, it neglects *exchange* effects (the antisymmetry of the electronic wavefunction, and the associated Fermi hole that this introduces around each electron). Next, it does not include the effect of electronic correlation – the Coulomb hole. Finally, it includes electron-electron self-interaction. In HF theory, the exchange and self-interaction error are corrected for simply by including the exchange integrals K_{ij}, but to recover the correlation effects, it is necessary to use very complex post-HF methods.

In DFT, the fourth term, or exchange-correlation functional, in principle contains all of the corrections needed to give the exact overall energy. E_{XC} is often split into one term corresponding to "exchange", E_X, and one corresponding to "correlation", E_C. This division into two terms reflects the fact that the neglect of exchange and self-interaction, on the one hand, and of correlation on the other, can at first sight be treated as separate effects.

Accordingly, the role of E_X is to correct for exchange and self-interaction effects. The interelectronic Coulomb repulsion energy, J_{ee}, is too large because it regards electrons of the same spin as approaching each other closely, whereas in fact the antisymmetry of the wavefunction keeps them apart. By assuming that the Fermi hole is relatively "small", it is possible to treat the exchange correction as an integral over all of space of a "local" effect. This local effect contributes a stabilisation term that is larger in regions of space where the density of electrons of a given spin is larger. In the simplest version of DFT, the *Local Density Approximation* (LDA), this term is given the following very simple expression derived from the Dirac formula which applies for a uniform "gas" of electrons:

$$E_X(\rho) = -2^{1/3}C\int\left(\rho_\alpha^{4/3} + \rho_\beta^{4/3}\right)dr \tag{5}$$

Here ρ_α and ρ_β are the densities of spin-up and -down electrons, respectively, at a given point in space. Like the Coulomb interelectronic repulsion term, this exchange functional does not evaluate to zero for one-electron systems, where there are of course *no* electron-electron interactions. In fact, the two terms are similar in magnitude in such cases, and roughly cancel out. In this sense, the exchange functional also corrects in part for self-interaction effects. For example, the LDA Coulombic energy J_{ee} of the hydrogen atom, calculated using a large basis set, is 0.2997 Hartrees, and the LDA E_X exchange functional is −0.2791 Hartrees. As can be seen, however, the two terms do not cancel out exactly, and this leads to the so-called *self-interaction error* of traditional DFT methods, which we will discuss below. These errors, associated with calculating the exchange energy from the local (or quasi-local, see below) density, suggest that DFT should perform much less well than HF theory, where exchange is treated exactly. In practice, however, this is not true. One might also think that a better form of DFT could be constructed, in which exchange is calculated *entirely* from the HF expression (not partly, as in hybrid DFTs), but correlation is incorporated using DFT expressions. Such approaches have been tried, but they do not perform well. This is partly because the HF description of exchange is not, in fact, exact: it is exact only within the approximation of a wavefunction represented by a single Slater Determinant. In DFT, the molecular orbitals and therefore the Slater Determinant used to expand the density are *not* assumed to correspond to the exact wavefunction (only to the exact density), so there is no reason to expect that they could be used to calculate exchange exactly. Also, exchange and correlation do not have exactly the same meaning in DFT and HF theory, and these subtle differences lead to problems in this type of approach.

The correlation part of the functional, E_C, is also derived, for simple functionals, from expressions derived for the uniform electron gas. As for the exchange term, the role of this part of the functional is to incorporate the stabilising effect of the Coulomb hole that surrounds electrons. Thus, in regions where there is a build-up of electron density – where the probability of finding *two* electrons simultaneously is larger – the Coulombic repulsion energy predicted from Eq. 4 will be an overestimate, as it neglects the correlation in electron motion. Again, like exchange energy, this effect is modelled in DFT as a *local* effect. Various different correlation functionals have been suggested, which are all given as integrals over space of functions of the density (and in some cases of its gradient). The mathematical form of these functionals can be found elsewhere (see [11] and other references below).

In the above discussion (in the context of HF theory), electron correlation was shown to play an important role in determining the relative energetics of spin states. In fact, within post-HF methods, the problem associated with calculating this energy difference is entirely due to the need to use more and more complicated wavefunctions so as to recover a significant proportion of the differential correlation effect between the two states. It might therefore be expected that, within DFT, the spin-state splitting would be determined mainly by the correlation functional. As will be seen below, however, this is not so, with the correlation functional playing little or no role in relative energetics. In simplistic terms, this can be seen as due to the fact that all correlation functionals make a similar contribution to the energy difference, whereas the different exchange functionals make different contributions.

In summary, the use of approximate functionals can lead to errors for several reasons: the neglect of correlation effects on electronic kinetic energy, the incorrect cancellation of the self-interaction involved in the Coulomb term, and the inaccurate treatment of exchange and correlation effects. It will be argued below that the errors associated with self-interaction and exchange play the most important role in calculations of spin-state energetics in TM compounds.

The simplest exchange-correlation functionals, derived from the density alone, give reasonable results for many molecular properties. However, it has been found that adding terms proportional to the local *gradient* of the density in both the exchange and correlation integrands leads to much more accurate results. Most modern DFT computations use these so-called "non-local" or gradient-corrected functionals.

It is useful to mention some of the popular DFT functionals that we will refer to below. Some of these are "complete", exchange and correlation functionals, such as the generalized gradient approximation of Perdew and Wang (abbreviated as PW91) [12] and the HCTH147 [13] and RPBE [14] functionals. Some include only an exchange functional and are designed to be used together with a correlation functional, such as Becke's 1988 functional (B or B88) [15] or the Gill 1996 (G or G96) expression [16]. The PW91 functional mentioned above in fact contains exchange and correlation contributions which can be used separately with other functionals. The most important correlation functionals are those of Perdew (P or P86) [17] and of Lee, Yang and Parr (LYP) [18].

The expressions given above for calculating the energy of a molecular system all use so-called *pure* density functional theory. That is, the energy is calculated from the density alone – although the kinetic energy is derived from a set of Kohn-Sham "orbitals". A very popular modification of DFT was suggested some years ago, in which the exchange energy is computed as a weighted sum of two terms, one of which is based upon the density alone, and the other is the Hartree-Fock exchange expression, using the Slater Determinant constructed from the Kohn-Sham orbitals. As discussed above, poor results are obtained when one uses this "exact" exchange expression on its own to calculate the exchange energy within DFT. However, as a mixture with DFT expressions, very accurate functionals can be obtained. The proportion of HF exchange was initially chosen to be 50% [19], yielding the BHLYP (or BHandHLYP) functional in conjunction with the B and LYP exchange and correlation functionals. In later work, the coefficient defining this proportion was *optimised* to reproduce computed energetics of a set of small main-group compounds as accurately as possible. This yields a family of functionals with 20% of exact exchange [20], by far the most common of which also includes contributions from the LDA exchange and correlation expressions, the B exchange functional, and LYP correlation. This B3LYP functional is one of the most accurate available at the present time [21]. It also performs very well for many species that do not belong to the initial set of compounds, including TM compounds, as shown by its countless applications in the field. There are many other so-called "hybrid" functionals, some of which, for instance B3PW91 or B3P86, use the same mathematical form and mixing coefficients as B3LYP but other correlation functionals. Others, such as B1LYP, PBE0, KMLYP, MPW1 K, use different proportions of exact exchange and different DFT expressions for exchange and correlation.

The development of new DFTs is a very active field of research, with new "pure" and hybrid functionals appearing all the time, some of which perform as well as or better than B3LYP for some (or all?) properties. In most cases, the researchers developing functionals test them on sets of compounds that do not contain any TM atoms. It then takes some time before such newly developed functionals become widely adopted by the broader computational community, and especially before they are used to try to predict spin-state splittings of TM compounds. Given the difficulties discussed below, it would seem desirable for TM compounds to be used earlier in the development phase of new functionals, although this is itself not straightforward given the need for reliable benchmarks.

3
Spin-State Splitting in Main-Group Compounds

Before addressing transition metal compounds, it is instructive to see how DFT compares to wavefunction methods for predicting the spin-state splitting of main-group open-shell compounds. Given the many calculations on this type of system in the literature, there is potentially a lot of data available pertaining to this issue. In line with the general high performance of the B3LYP functional, it often seems to give good results for state splittings, for example of carbenes [22]. However, other functionals also seem to perform well [23], suggesting that there may not be great differences in this field between them. Perhaps because of this good agreement, and the general good behaviour of standard functionals such as B3LYP, there do not appear to be many systematic studies of DFT-dependence. For this reason, we include here, as a reference for the following sections on transition metal compounds, state splittings calculated for two simple compounds with a broad range of functionals [24].

Table 1 shows the predicted splitting between the triplet ground state and the closed-shell first singlet excited state of the methylene biradical, CH_2. To take the ab initio results first, it can be seen that HF exaggerates the stability of the triplet, for the reasons discussed above. Next, the simplest correlated method, MP2, does not improve the result much, mainly due to near-degeneracy effects in the singlet state. All of the multi-reference methods, including the very simple two configuration CASSCF(2,2) method, give results in close agreement and close to the exact value. The highly-correlated single-reference CCSD and especially CCSD(T) methods also perform well.

Table 1 DFT and ab initio calculated energy splittings (in kcal/mol) between the triplet and singlet states of CH_2. The indicated value does not include a correction for zero-point energy

DFT/6-311G** (optimized)		Ab initio/cc-pVQZ (at B3LYP/6-311G** geometry)	
DFT	$\Delta E(1,3)$	Ab initio level	$\Delta E(1,3)$
LDA	13.43	HF	24.98
PW91	15.46	MP2	19.50
BLYP	10.26	CCSD	10.39
BP86	13.64	CCSD(T)	9.52
B3PW91	15.94	CASSCF(2,2)	10.49
B3LYP	11.39	CASSCF(6,6)	10.15
B3P86	14.60	MRCISD	9.15
B1LYP	11.57	MRCISD+Q	8.99
BHandHLYP	12.66	MRACPF	9.04
HF	26.02	MRAQCC	9.09

The most important trend within the DFT results is that there is *no* systematic dependence of $\Delta E(1,3)$ on the type of exchange functional used. In fact, the correlation functional appears to be more important: The PW91 and B3PW91 functionals give very similar state-splittings which are clearly too negative, BP86 and B3P86 also overestimate the splitting, whereas BLYP, B3LYP and B1LYP all give values which are fairly close to experiment. Of the functionals using the LYP correlation functional, only BHandHLYP, with its very high proportion of exact exchange (50%), overstabilises the high-spin state to a significant extent. Finally, even the simplest LDA functional gives a reasonable value, which is not out of line with those obtained with the other functionals.

This pattern seems to be fairly common for main group compounds. In Table 2, results with the same range of functionals are shown for the singlet-triplet splitting in methyl-silylene [24]. In this system, the unpaired electrons reside on a different atom (Si versus C), and the singlet is now the

Table 2 DFT and ab initio calculated energy splittings (in kcal/mol) between the triplet and singlet states of CH_3SiH. HF and DFT results involve full geometry optimization with the 6-311G** basis; CCSD and CCSD(T) values use the cc-pVQZ basis at the B3LYP/6-311G** geometry; no corrections for zero-point energy have been included

Method	$\Delta E(1,3)$	Method	$\Delta E(1,3)$
LDA	−23.15	B3P86	−20.00
PW91	−19.20	B1LYP	−22.78
BLYP	−23.68	BHandHLYP	−21.97
BP86	−20.59	HF	−7.67
B3PW91	−18.78	CCSD	−22.66
B3LYP	−22.96	CCSD(T)	−23.17

ground state ($\Delta E(1,3)$ is positive). Using the CCSD(T) results as a reference, it can be seen that the functionals using the LYP functional are again most accurate, with the PW91 and, to a lesser extent, P86 correlation functionals leading to an over-stable triplet (or under-stable singlet). As in the case of CH_2, HF clearly favours the triplet, the LDA performs well, and even the largest differences between functionals are fairly modest.

In summary, most DFT functionals seem to reproduce spin-state splittings of main-group compounds fairly accurately. The differences between functionals, from experiment or accurate computation, are typically below 5 kcal/mol, and the nature of the correlation functional seems to be somewhat more important than that of the exchange functional. As expected from its general good performance for predicting thermochemistry of main-group compounds, the functional chosen by default in many computational studies, B3LYP, also gives good results for spin-state splittings.

4 Spin-State Splitting in TM Compounds

The situation for transition metal compounds is rather different from that just discussed for main-group species. Different functionals frequently predict spin-state splittings which differ by 10 kcal/mol or more, and this has been noticed and reported by many different groups. A lot of effort has been put into ascertaining which functional gives the best results for a given system. This is not always easy, as an accurate experimental or computational calibration is not always available. An additional difficulty comes from the fact that many computational studies of TM compounds use relatively small basis sets, such as the double zeta, unpolarized, LanL2DZ ECP basis. Although DFT is not as demanding in terms of basis sets as correlated ab initio methods (only the one-particle space of the Kohn-Sham orbitals needs to be spanned, not the two-particle inter-electronic cusp, the accurate description of which requires very large basis sets), and some properties calculated with basis sets as small as this can be more or less converged, it is frequently the case that basis set effects of 10 kcal/mol can affect such results. We will now review the results of calculations on a range of TM compounds.

4.1 Atoms and Atomic Ions

It might appear at first sight that atomic systems are ideal for testing and benchmarking the performance of DFT in predicting spin-state splitting. The energies of the electronic states of these species are known very accurately from experiment, they are amenable to accurate computation using ab initio methods, and a broad range of *d* orbital occupation patterns and spin multiplicities are covered if one considers all of the transition metals. However, there are some problems associated with these species. First and foremost, atomic states tend to be degenerate due to orbital angular momentum as well as spin. The simple-minded approach to representing these states is to carry out DFT calculations with the electrons assigned to orbitals in such a way as to obtain the correct overall occupation pattern $nd^q(n+1)s^r$ – for example the valence electrons in the iron atom will be placed so as to obtain a $3d^6 4s^2$ configuration consistent with the ^5D ground state of the atom. However, the angular degeneracy of the atomic state means that it can, in principle, be represented in several different ways in terms of the occupations of the individual *d* orbitals: $(z^2)^2(xy)^1(xz)^1(yz)^1(x^2-y^2)^1$ or $(z^2)^1(xy)^2(xz)^1$ $(yz)^1(x^2-y^2)^1$, and so on. With many functionals, these different occupation patterns generate significantly (several kcal/mol) *different* energies. It has been proposed that either carefully chosen single configurations or weighted averages over several configurations should be used as atomic multiplet energies. For a discussion of these issues, see [25–27] and references therein. Obviously, the calculated spin-state splitting will depend on whether or not these more sophisticated approaches are used or whether one has simply entered the required charge and multiplicity into the electronic structure software and hoped for the best!

Another objection to the use of atoms as reference systems for calibration is that their electronic structure is often not typical of that of the metallic complexes for which prediction of spin states is desirable. For example, the $(n+1)s$ orbital of a metal in a complex is usually not occupied, whereas it is in bare metal atoms. For this reason, unless there are good grounds for thinking that the behaviour of functionals should be fairly constant whatever the electronic structure and bonding environment, then the "best" functionals for atoms may perform poorly for complexes. We note that as well as atoms and atomic ions, small, highly unsaturated gas phase molecules such as metal monohydrides MH, monoxides MO, and monocarbonyls MCO can be argued to have somewhat unrepresentative electronic structures. For this reason, it is advisable to avoid extrapolating too confidently from computational studies of these systems to larger complexes.

Of course, in some cases it is important to be able to calculate the energies of atoms, and it is also interesting to examine how functionals behave for these systems. Thus, despite the problems with these calculations, many studies have been carried out on atoms and atomic ions, often with the specific aim of testing functional behaviour. To take just one example, Markovits et al studied the ground state of the first row transition metal atoms using the PW91 and B3LYP functionals as a part of their study of the absorption of metal atoms on magnesium oxide surfaces [28]. For most metals, the lowest configuration was found to correspond to the experimental ground state. For V, however, PW91 predicts a $3d^4 4s^1$ quintet ground state, whereas B3LYP agrees with experiment in finding a $3d^3 4s^2$ triplet ground state. For

Ni, both functionals predict a triplet ground state, but with PW91, this is a $3d^94s^1$ configuration, instead of $3d^84s^2$ as found experimentally and with B3LYP. This example illustrates the second point made above: the different states involved differ in their 3*d* occupation instead of, or as well as, in their spin state. In metal complexes, competing spin states differ instead in terms of the spin coupling of the n*d* electrons.

4.2 Organometallic and Coordination Complexes

After these highly unsaturated species, the rest of the review will focus on more realistic, multi-ligand complexes. It is useful to make a distinction between organometallic and Werner-type coordination compounds. Organometallic compounds typically involve "strong field" ligands such as carbonyl, whose orbitals interact strongly with the *d* orbitals of the metal through both bonding and back-bonding interactions. This is also favoured by the fact that in many organometallic compounds, the metal atom has a fairly low formal oxidation state, and indeed is fairly electron rich, so that its occupied *d* orbitals are fairly high in energy, and interact significantly with the ligand orbitals. All of this results in fairly strong splitting of the formal *d* orbital energy levels, which results in a significant preference of organometallic compounds for closed-shell singlet electronic ground states. For example, the success of the 18-electron rule in rationalising bonding in these compounds clearly correlates with this preference for singlet states.

In contrast to this, coordination compounds typically involve weaker-field nitrogen, halogen or oxygen-based ligands on the metal centre. Also, the metal often has a fairly high formal charge of +2 or +3, and so has quite small *d* orbitals spatially, which overlap less with ligand orbitals. As a result, *d* orbital splitting is less pronounced, and both high-spin and low-spin electronic states can be observed. For example, octahedral Fe(II) complexes intrinsically have a formal electron count of 18, yet both quintet and singlet (and indeed sometimes triplet) ground states are very common for these species. For these compounds, then, the occurrence of close-lying spin states is a general rule, and being able to predict the ground state is crucial if one wants to be able to predict such things as geometrical structure or spectroscopy.

Of course, for many TM species, it can be difficult (and conceptually not very useful!) to decide whether they are "organometallic" or "coordination" compounds – for many complexes, a good case could be made for them belonging to both groups (or neither). Also, the hand-waving rules given above whereby organometallic compounds tend to have low-spin and indeed most often singlet ground states, whereas coordination compounds tend to have close-lying high- and low-spin states, are much too simple for them not to fail quite frequently. From our present point of view, it is important to point out that there are many stable organometallic species, and even more reactive intermediates, in which an electron count lower than 18, an odd number of electrons, or the presence of comparatively weak-field ligands, leads to the existence of low-lying open-shell electronic states. For these cases, it is important to be able to predict spin-state splittings accurately, as it is more generally for coordination compounds.

4.2.1 Iron Tetracarbonyl

It is instructive to start our overview of TM compounds with a very simple species, which has been the object of many computational studies: iron tetracarbonyl. As will be seen, many of the trends concerning DFT and spin-state splitting are observed for this compound. $Fe(CO)_4$ is a 16-electron reactive intermediate which has been shown to have a triplet ground state in the gas phase, but with a low-lying singlet state just above. Both electronic states have C_{2V} geometries, which can be thought of as arising from distortion of the species obtained by removing an equatorial carbonyl from trigonal bipyramidic $Fe(CO)_5$ (See Fig. 4). The ground state has three doubly-occupied orbitals which are mainly metal-3*d* in character, and two singly occupied orbitals of b_1 and a_1 symmetry, respectively. The first of these is essentially the metal $3d_{xz}$ orbital – the *d* orbital lying in the "equatorial" plane. The second is mainly the iron d_{z^2} orbital, with some 4*s* character, and is oriented along the C_{2V} axis – which corresponds to what was, in formal terms, the broken Fe–C bond in $Fe(CO)_5$. In the singlet, the b_1 orbital is doubly occupied, and the a_1 orbital is empty. This occupation pattern is fairly typical of other unsaturated 16-electron singlet/triplet pairs.

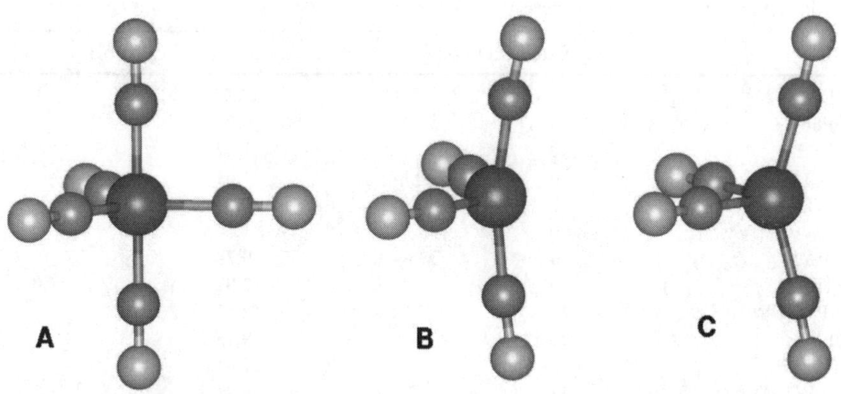

Fig. 4 Three-dimensional representation of the structures of A: $Fe(CO)_5$; B: $^1Fe(CO)_5$, and; C: $^3Fe(CO)_4$; showing how the latter two structures can be derived from the first one by removing one CO

For the reasons discussed above, early ab initio calculations [29, 30] tend to exaggerate the stability of the triplet ground state, even when using correlated wavefunctions, as it is difficult to recover all of the differential correlation effects. The computations thereby find a large (~15 kcal/mol or more) splitting between the triplet and singlet states. There have been many DFT studies of this species [31–35]. The geometries found in these different studies are generally in good agreement, but the spin-state splitting is found to vary quite strongly depending on the functional used. Thus, "pure" DFT methods give rather low excitation energies, for example ΔE=0.5 kcal/mol with BP86 and BLYP [32], whereas hybrid functionals give larger splittings, like 8.7 kcal/mol with B3LYP [32].

In a recent study, we calculated the state splitting of $Fe(CO)_4$ using a variety of DFT and wavefunction methods [36]. The DFT calculations yielded results very similar to those found above, except that we surveyed a broader range of functionals (see Table 3). This reveals some interesting points. First, the splitting does not strongly depend on the nature of the correlation functional, for a given exchange functional (or mixture of functionals). Thus BP86, BLYP, and BPW91 give very similar values. Second, there are *enormous* differences in the computed ΔE upon varying the exchange functional, and especially the amount of exact exchange. Thus, all pure functionals give a very small splitting (LDA is a slight exception, for which the *singlet* is the ground state, probably due to the absence of gradient corrections), functionals with 20–25% HF exchange give values close to 10 kcal/mol, and the two functionals KMLYP [37] and MPW1 K [38], developed to give accurate bar-

Table 3 DFT spin-state splitting $\Delta E(1,3)$ between $^3B_1$ and $^1A_1$ $Fe(CO)_4$, in kcal/mol. The negative value for the LDA means that the singlet is found to be the ground state with that method. Also shown is the proportion of "exact" exchange used in the functional. From [36]; includes a correction for zero-point energy

DFT	Exact exchange (%)	$\Delta E(1,3)$ (kcal/mol)
LDA	0	−5.91
BP86	0	0.92
BLYP	0	1.43
G96LYP	0	1.61
HCTH147	0	0.43
BPW91	0	1.64
B3LYP	20	9.78
B1LYP	25	12.92
mPW1PW91	25	12.11
PBE0	25	9.15
KMLYP	55	21.61
MPW1 K	42	20.21
B3PW91	20	9.65
B3PW91*	15	6.88
B3"PW91**	10	4.09

riers for some small gas-phase reactions, and containing ~50% HF exchange, give splittings close to 20 kcal/mol. Whilst not fully conclusive, our wavefunction-based calculations (CASPT2 and CCSD(T) with large basis sets) agreed that the correct splitting should be somewhere between the value given by the pure functionals and that given by B3LYP.

There is an interesting exception to the rule that pure functionals predict near-degenerate singlet and triplet states. The modified form of the "pure" PBE functional described by Hammer et al [14], called RPBE, has been reported [35] to predict a triplet-singlet energy difference for $Fe(CO)_4$ of 4.0 kcal/mol, which is probably much closer to the correct value than that computed with any other "pure" functional. Whilst the RPBE functional has been very enthusiastically adopted by the periodic DFT community for calculations on surfaces – the area for which it was developed – very few applications have been reported for molecular TM complexes, so it is not yet clear whether the good result for $Fe(CO)_4$ will be transposable to other systems.

In the last three rows of Table 3, the effect of varying the proportion of exact exchange whilst keeping all other aspects of the functional constant shows that the spin-state splitting does indeed depend on the proportion of exact exchange. We found that the modified B3PW91* functional, with a reduced 15% contribution of exact exchange, gave the closest agreement with what we believe to be our "best" wavefunction-based predictions for the spin-state splitting and the bond energy in $Fe(CO)_5$ (which is not discussed here). The dependence on the proportion of exact exchange is shown more clearly in Fig. 5, where we have followed the example of Reiher et al [39]

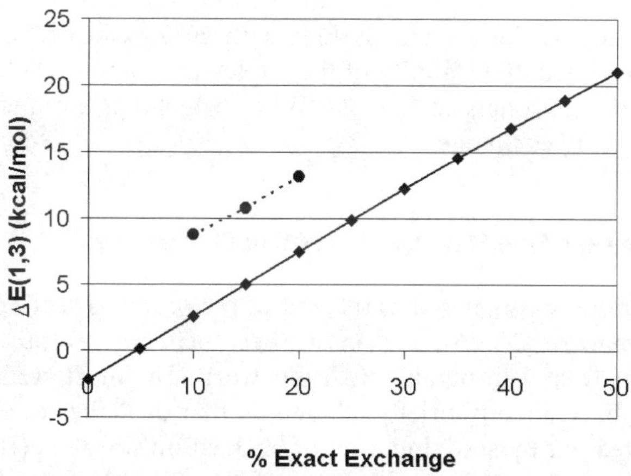

Fig. 5 Computed [23] spin-state splitting $\Delta E(1,3)$ between $^3Fe(CO)_4$ and $^1Fe(CO)_4$, as a function of the proportion of exact exchange. Diamonds and solid line: B3LYP(c_3) results; discs and dotted line: B3(c_3) results (only the exchange parts of the DFT functional were used)

(see below), and plotted ΔE for 11 different B3LYP(c_3) functionals, in which the c_3 coefficient defining the percentage of exact exchange is varied from 0–50% [24]. As can be seen, the dependence is almost linear, with increasing amounts of exact exchange progressively favouring the triplet more and more.

A final, and remarkable, piece of evidence relating to $Fe(CO)_4$ and showing the importance of the exchange functional in establishing the spin-state splitting is also shown in the Figure. We have calculated the spin-state energy difference using modified B3LYP functionals, in which (a) the proportion of exact exchange was varied, and (b) the local and non-local correlation functionals were *completely removed* – only exchange contributions were included. These calculations actually give spin-state splittings which are only slightly more in favour of the triplet than with the full B3LYP(c_3) functionals. Like the latter, though, the predicted energy difference is found to vary significantly depending on the proportion of exact exchange. Clearly, the meaning of "exchange" in DFT is quite different from that which it has in HF and post-HF theories: the HF spin-state splitting in this system is ~38 kcal/mol [24], so the much lower values found for these "exchange only–B3LYP" functionals means that the exchange functional must introduce a description of some of what would be called correlation effects within wavefunction theory.

In summary, $Fe(CO)_4$ has a triplet ground state, with a closed-shell singlet lying slightly higher. The exact state splitting is unknown, but is likely to be between 4 and 8 kcal/mol. Pure DFT functionals (except RPBE) predict a state-splitting which is clearly too small, with the singlet lying ~1 kcal/mol above the triplet. Hybrid functionals predict a splitting which increases almost linearly with the proportion of mixed-in exact exchange. In particular, functionals such as the popular B3LYP, with 20% exact exchange, seem to slightly overestimate the stability of the triplet, and better results seem to be obtained with functionals such as B3PW91*, which has a reduced 15% contribution from HF exchange.

4.2.2 Exact Exchange and Spin-State Splitting: Other Observations

The observations summarised at the end of the previous section, and drawn from our own work [24, 36], have in fact been made by several other authors both previous to and in parallel with our work. The most systematic study, to which we have already briefly alluded, is that of Reiher et al [39]. These authors started out by studying a set of bridged dithiolate Fe(II) complexes, shown in Fig. 6. The spin multiplicity of the ground state of these compounds was known experimentally. The authors found that use of the BP86 functional wrongly predicted that some high-spin (S=2) compounds should have a lower-lying low-spin (S=0) state, whereas use of the B3LYP functional

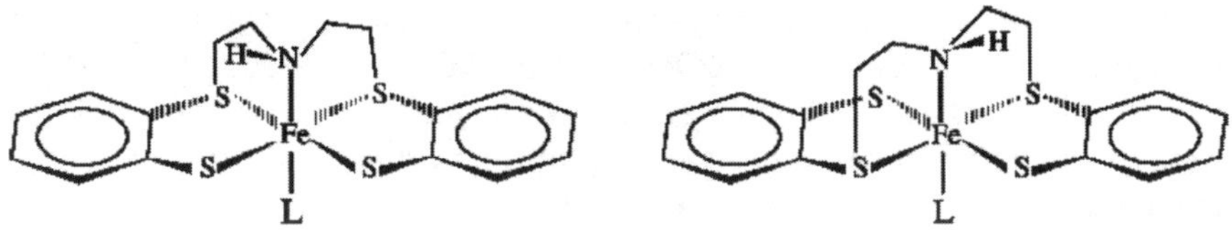

1 (L)
L=NH_3 [50], N_2H_4 [50]

2(L)
L=CO [51], NO^+ [52], PR_3 [4]

Fig. 6 Fe(II) compounds studied by Reiher et al [39]. Reproduced with permission from [39]

led to the opposite problem: some low-spin species were predicted to have high-spin ground states.

The authors went on [39] to study the effect on the spin-state splitting for each of the compounds of using a range of hybrid functionals based on B3LYP but with a systematically varied proportion of exact exchange. They found an almost exactly linear dependence, as shown in Fig. 7. Note that the slope of the lines relating spin-splitting to exchange is rather large – changes in relative energetics of more than 25 kcal/mol (100 kJ/mol) are observed upon going from pure functionals to those with 20% of exact exchange. This effect is clearly *not* a small one that can be ignored in all but the most quantitatively-focused studies! Values of between 10–15% define a so-called “optimum” region, within which the correct ground-state is predicted for all the compounds studied. With either lower or higher amounts of exchange, at least some of the predicted ground states were wrong. Accordingly, the au-

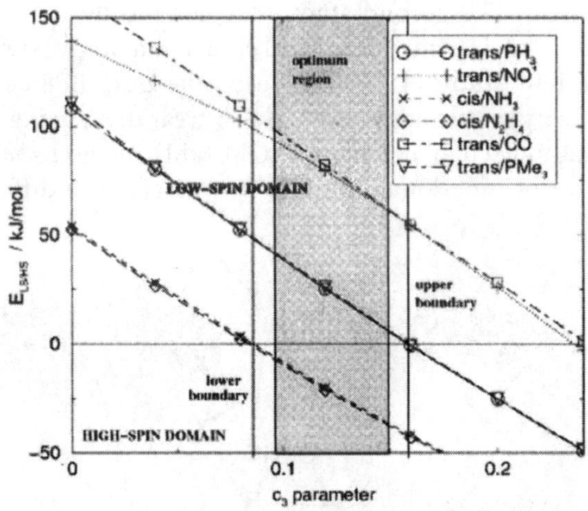

Fig. 7 Computed spin-state splittings ELS/HS for the compounds of Fig. 6 [39], as a function of the proportion of exact exchange. Reproduced with permission from [39]

thors proposed [39] the use of a modified B3LYP functional, with 15% exact exchange (c_3=0.15 instead of 0.20), which they refer to as B3LYP*.

In subsequent work, Reiher et al have tested the performance of this B3LYP* functional much more widely [40]. First, they found that the agreement with experiment for the set of atomization and ionization energies and of electron affinities for the main-group molecules contained in the G2 test set is not appreciably worse with B3LYP* compared to B3LYP. Next, they explored the dependence of spin-state splitting on the c_3 coefficient for different spin states of metallocenes $M(C_5H_5)_2$ (M=Ti, V, Cr, Mn, Fe, Co, Ni) and of bis-benzene complexes, $M(C_6H_6)_2$ (M=Ti, V, Cr, Fe, Co), $Mn(C_6H_6)_2^+$ and $Ni(C_6H_6)_2^{2+}$. As for the Fe(II) complexes discussed in their previous paper, most of these species showed a relative stabilization of the high-spin state upon increasing the amount of exact exchange. The authors draw particular attention to the cases of manganocene ($Mn(C_5H_5)_2$) and cobaltocene ($Co(C_5H_5)_2$). Experimentally, the first of these has a sextet ground state whereas the second is a doublet. Agreement with experiment is only found for c_3 values larger than 14% and smaller than 22%. This interval overlaps in a small region with that found in their earlier study, and reinforces their recommendation that B3LYP* should be a more accurate functional for describing spin-state splitting than B3LYP itself, and is much more accurate than "pure" density functionals.

This observation is informally supported by work from other groups. For example, the mechanism of the *bis*(imino)pyridyl iron(II) catalyzed polymerization of alkenes has been of considerable interest, and two major computational studies have appeared on this topic [41, 42]. Both studies have addressed parts of the singlet, triplet, and quintet potential energy surfaces relevant to the individual steps in propagation and termination. Figure 8 shows the catalyst system. The second study [42] has mostly used the B3LYP hybrid functional, and includes a critical and comparative review of differences with the first study [41], which used the pure BP86 functional. The studies address large molecules and to do so treat them using a mixed quantum-mechanical/molecular mechanics (QM/MM) method. As well as differing in the choice of functional, the two studies also use different computer

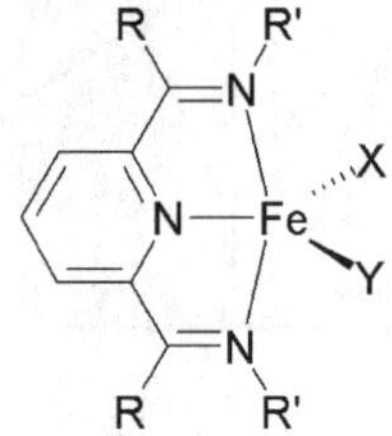

Fig. 8 Generic structure of the *bis*(imino)pyridyl iron(II) catalyst system studied in [41, 42]

programs, and different MM force-fields, so that straightforward comparison is not possible. Nevertheless, spin-state splitting for a number of intermediates are systematically and significantly (10 kcal/mol) more in favour of the high-spin state in the B3LYP study than in the BP86 study. In the supplementary material of [42], the authors report the results of calculations using both functionals, with the same basis set, for a small, QM-only, model dichloride system (Fig. 8, R=R′=H, X=Y=Cl). This enables a consistent comparison of energetics. BP86 predicts that quintet and triplet should be degenerate, with the singlet 3 kcal/mol higher. B3LYP gives relative energies of 0.0, 13.3 and 21.3 for quintet, triplet and singlet. As the full dichloride species (R′=bulky aryl group) is experimentally known to be a quintet, the authors note that B3LYP appears to describe spin-state energetics in this system more accurately than BP86.

In a related study on cobalt(II) *bis*(imino)pyridyl catalyzed polymerization [43], Margl et al have likewise compared the behaviour of BP86 and B3LYP for a number of species on the catalytic cycle. Their results (also presented in the Supplementary Material of their paper [43]) show that the spin-state splitting between doublet and quartet states is ~25 kcal/mol more in favour of the quartet with the hybrid B3LYP functional than with BP86. As in the iron case, the B3LYP result is in somewhat better agreement with experiment than the BP86 one, with respect to spin-state splitting, as it correctly predicts the quartet ground state of the catalyst precursor, whereas BP86 predicts a doublet ground state.

In brief, there is lots of other evidence in the literature that hybrid functionals favour high-spin states:

- Ghosh and Taylor have calculated a 4 kcal/mol lower energy difference with BP86 than with B3LYP between the sextet ground state and lowest quartet excited state of the hexa-solvated iron(III) ion, $Fe(H_2O)_6{}^{3+}$ [44].
- Scherlis and Estrin [45] have compared HF, BP86, BLYP, B3LYP, and BHandHLYP calculations on iron atoms and iron-porphyrin systems, and found that the pure functionals overestimate the stability of low-spin states, whereas HF favours the high-spin states.
- Several groups have studied the energy difference between low-spin and high-spin forms of so-called "spin-crossover" compounds such as $Fe(phen)_2(NCS)_2$ [46, 47], $Fe(tpen)^{2+}$ (tpen=tetrakis(2-pyridylmethyl)-ethylenediamine) [48, 49], and others [46], using different functionals. The energy difference is best described by B3LYP* [47], then by hybrid functionals, with pure functionals favouring the low-spin form.
- The energetics of the low-, intermediate-, and high-spin states of the key species on the potential energy surface for oxidation of alkenes by Mn(salen) (and the related chromium species [50]) complexes have received considerable attention, following the suggestion [51] by Linde et al that spin-state changes play an important role in the diastereoselectivity of this

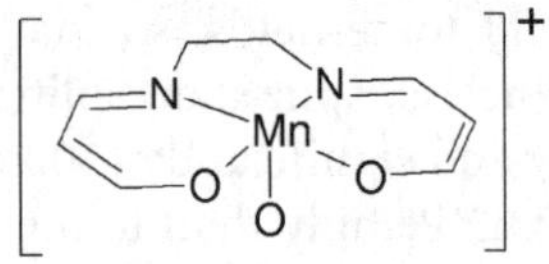

Fig. 9 Model system used to study spin-state splittings and reactivity in Mn(salen) oxidation

important synthetic transformation. The key manganese oxo intermediate, formally with an Mn(V) centre, has close-lying singlet, triplet and quintet states [52–54]. Calculations on a simple model compound, shown in Fig. 9, lead to very different relative energetics for these states. Thus, most functionals predict similar energies for the singlet and triplet states, but hybrid functionals place the quintet state close in energy as well, whereas with pure functionals it lies ~15 kcal/mol higher.

For this system, there is a complicating factor which is not always properly taken into account when discussing spin-state energetics: the manganese oxo species does *not* really belong to type A in the classification shown in Fig. 1, because the unpaired electrons in the different spin states are not purely located on the metal. Significant spin-contamination in the triplet [53] is inconsistent with it being described purely as a Mn(V) centre with two unpaired electrons, bearing an oxo (−2) ligand. Instead, it must be described as a mixture between this "oxo" structure, and an "oxy" resonance form. The latter involves a high-spin Mn(IV) centre, with three unpaired electrons contributing an overall 3/2 spin angular momentum, low-spin coupled to an oxy radical doublet. In this picture, there are four unpaired electrons, one of which is not on the metal! The quintet corresponds to a high-spin coupling of the Mn(IV) and oxy radical spins. The change from singlet to triplet therefore involves some redox changes as well as a change in spin on the metal, and the change from triplet to quintet involves some redox change, and a change in the relative alignment of the spin contributed by two different parts of the molecule – as in case C of Fig. 1. This is a very different situation to that found in systems such as the Fe(II) or Co(II) complexes, in which the different spin states do mostly arise as a consequence of different spin-pairing on the metal centre itself. It is not clear that spin-state splittings in borderline case C systems will follow the same trends as normal case A compounds.

- In their study of the spin-forbidden addition reactions of CO and N_2 to the 16-electron spin-triplet $Cp^*MoCl(PMe_3)_2$, Jensen and Poli calculated the energy difference between the triplet ground state and lowest singlet excited state of the fragment using both B3LYP and BPW91 functionals, as well as various wavefunction methods [55]. The HF method gave a large splitting,

of 19.5 kcal/mol, B3LYP gave a much reduced energy difference of 5.6 kcal/mol, and the pure functional gave a still smaller 2.9 kcal/mol.
- As a part of their attempt to evaluate the capability of the PM3 semiempirical method to correctly predict the correct spin state of technetium complexes, Ball et al [56] compared spin-state splittings computed for several of the target compounds with four different functionals, BP86, BLYP, B3P86 and B3LYP. Singlet-triplet or doublet-quartet splittings were typically 3–5 kcal/mol more in favour of the singlet with the pure functionals than with the hybrid ones, with still larger effects for the singlet-quintet or doublet-sextet energy differences.

It is perhaps worth noting in the context of these last two examples that the effect of exact exchange on spin-state splitting is apparently much smaller for complexes of the second-row TM elements. The cobalt and iron examples given earlier lead to differences of up to 20 kcal/mol or more between spin-state splittings computed using standard pure (BP86) and hybrid (B3LYP) functionals. The differences in the case of the molybdenum and technetium compounds are much smaller. This is not limited to early second-row elements or indeed to the second row, as shown in Table 4, in which we report computed spin-state splitting for the second- and third-row analogues of $Fe(CO)_4$, $Ru(CO)_4$ and $Os(CO)_4$, obtained with a similar basis set and a similar range of functionals to that used for $Fe(CO)_4$ (Table 3, [36]). Note that spin-orbit coupling – not included in the computations – may affect the energy of the "triplet" state of these heavy species, especially the Osmium complex.

These 16-electron species adopt rather similar geometries to the corresponding spin-states of $Fe(CO)_4$, but unlike for the latter, the singlet is predicted to be the ground state at all levels (except HF for $Os(CO)_4$). This is

Table 4 DFT spin-state splitting $\Delta E(1,3)$ between $^3B_1$ and $^1A_1$ states of $Ru(CO)_4$ and $Os(CO)_4$, in kcal/mol; the negative values correspond to singlet ground states (from [24]). Calculated using a 6-31G* basis on C and O, LACV3P on Ru or Os, at the B3LYP/6-31G* geometry. Does not include a correction for zero-point energy

DFT	$\Delta E(1,3)$ Ru	$\Delta E(1,3)$ Os
LDA	−21.4	−17.8
BP86	−17.7	−15.9
BLYP	−18.5	−17.7
PW91	−17.9	−15.7
B3LYP	−16.7	−14.7
B3PW91	−15.6	−12.6
B1LYP	−15.8	−13.8
BHandHLYP	−13.0	−9.8
KMLYP	−14.0	−9.0
HF	−0.3	5.8

basically due to the larger spatial extent of the 4*d* and 5*d* orbitals, and hence to better bonding interactions with the ligands, a large *d* orbital splitting, and decreased *d*-*d* Coulombic repulsion. As can be seen in the Table, for these species, the computed spin-state splitting varies by only 5.5 and 6.1 kcal/mol, respectively, between the pure BLYP functional and the BHandHLYP functional, which includes a very large amount, 50%, of exact exchange. For a similar pair of functionals, BLYP and KMLYP, the spin-state splitting of $Fe(CO)_4$ changes by over 20 kcal/mol. Nevertheless, the overall trend remains that the high-spin state is favoured upon mixing in increasing amounts of HF exchange.

5 Discussion

The above examples show that there is abundant evidence in the literature that mixing in of HF exact exchange can lead to dramatic changes in spin-state splittings of TM complexes as compared to results obtained using pure functionals. The energy effect depends approximately linearly upon the proportion of exact exchange, and leads to the stabilisation of the high-spin state (or destabilisation of the low-spin state). It is particularly strong for first-row transition metals.

The most usual interpretation provided for this effect is implicit in our introduction on HF and DFT methods: HF accounts for the Fermi hole correlation effect between electrons of same spin, but does not describe the more general (Coulomb hole-derived) correlation between electron motion. It therefore over-stabilises high-spin states. In contrast, pure DFT often over-stabilises low-spin states. Note that the reason for this is not clear, and none is usually given when this point is being discussed, but it is a fairly common observation. Taken together, these two trends mean that mixing in HF exchange terms in the DFT functional will lead to error compensation, or to a more accurate functional, depending on the point of view. From the pragmatic point of view, this point of view seems acceptable at first sight: if, for a given compound, spin-state splitting depends strongly on exact exchange admixture, and pure functionals overstabilise the low-spin form whereas HF overestimates the high-spin form, then adjusting the amount of exact exchange should enable parameterisation of a functional which gives good agreement with experiment for that system.

However, this account does raise a number of issues, concerning transferability (or universality), calibration, the role of exchange functionals in pure DFT, and the self-interaction error.

Transferability is in some ways a double problem. The first of these relates to whether an optimised exact exchange contribution able to reproduce spin-state splittings in a given compound or class of compounds will reproduce them equally well for other compounds. The second is whether the preferred value for the exact exchange contribution is able to give accurate predictions for other properties.

Taking the second problem first, Reiher et al [40] have found that changing the proportion of exact exchange from 0.20 to 0.15, whilst greatly improving the predicted spin-state splittings for some compounds, has only a small effect on the computed thermochemistry of main group compounds. The resulting B3LYP* functional yields results on the G2 thermochemical test set which are of very similar quality to those obtained with B3LYP itself. In other words, energetics for many main-group compounds are sufficiently insensitive to exact exchange that this can be modified slightly in order to improve behaviour for TM spin-state splitting. However, large changes are impossible because this would lead to inferior results – the mixing ratio was *optimised* to 20% by fitting to thermochemical data for a large number of main group compounds. It is the consequent superior thermochemical behaviour compared to pure functionals which is, after all, the main reason why B3LYP has become so popular.

Speaking more generally, though, it does not appear that all properties *are* well computed by functionals with a proportion of exact exchange similar to B3LYP or B3LYP*. In transition metals, it has been shown [57] that both pure functionals and hybrid functionals with a moderate amount of exact exchange, such as B3LYP, provide a description of bonding in Cu(II) complexes such as $CuCl_4^-$ which is too covalent. The authors of this study found that increasing the admixture of exact exchange to 38% produced a better account of the bonding. Equally, the barrier heights of many simple reactions are poorly described by B3LYP, and it is found empirically that better agreement with experiment is obtained only upon mixing in very high (~50%) proportions of exact exchange (see for example [37, 38]). As shown in Tables 3 and 4, and elsewhere by Reiher et al [39, 40], such HF exchange-rich functionals give very poor spin-state splittings. Other researchers have varied the degree of exact exchange and studied other properties. For example, Tozer et al find that a c_3 coefficient of between 0.2 and 0.3 yields the best agreement with experiment for the prediction of magnetic resonance shielding tensors [58]. In fact, as discussed below, there is no real reason why there should be a single "best" value of the coefficient for exact exchange admixture, which yields perfect results for all properties, and the empirical evidence suggests that there is indeed no such value.

In fact, we believe there is some evidence that spin-state splittings, in at least some cases, are reproduced best by functionals with a very different proportion of exact exchange. Thus, in the CpCo(CO) 16-electron system, all ab initio methods and DFT functionals predict a triplet ground state, with a singlet excited state [59]. In recent calculations on this system [60], we have found a strong dependence of the spin-state splitting on the proportion of exact exchange included in the functional. Thus, with the B3LYP functional, the difference is of 25.7 kcal/mol, whereas with BP86, the splitting is as small as 14.9 kcal/mol. We have carried out CCSD(T) calculations on this system, using very large basis sets on the cobalt atom (quadruple zeta, with 3*f*, 2*g* and 1*h* polarization functions). At this level, the splitting is of 15.4 kcal/mol – much closer to the BP86 value than to the B3LYP one. In this system, if the CCSD(T) result is reliable, then the optimum exchange energy mixing is clearly of the order of 0%, not 20 or indeed 15%. That being said, the accuracy of single-reference treatments of first-row transition metal systems is much disputed. In some recent work [36], we have argued that these methods should in fact be quite accurate, even when, as here, fairly large t_1 diagnostic values [61] of ~0.05 are obtained. However, given that this assessment of the reliability of the CCSD(T) method is still the subject of much discussion (see [62] for instance), and given that this appears to be one of the only cases where pure functionals appear to behave better than those with significant admixture of exact exchange, it is maybe prudent to reserve judgement until more evidence is available.

The previous paragraph serves to highlight a very important point for the whole area of prediction of spin-state splittings in TM compounds, that of calibration: to reach some level of confidence that a given functional predicts spin-state splittings accurately, one needs to know what the true splittings are. This is not as straightforward as it sounds. First, for main-group systems containing relatively few atoms, it is possible to refer DFT calculations to the results obtained from highly accurate *computational* calibration. For example, composite calculations using methods such as the popular G2 or G3 approaches [63, 64], or simply very large basis-set CCSD(T) or MR-CI computations, can be carried out in such a way as to reach chemical (1 kcal/mol) accuracy. The above discussion of our CCSD(T) computations on CpCo(CO) [60] shows that reaching that level of accuracy is still elusive for many TM compounds. As well as all of the intrinsic difficulties in accounting for the complex electronic structure of *d* elements, this is due to the fact that even small TM compounds are quite large by the standards of accurate ab initio computations if, that is, they are realistic in the sense of having a near-complete coordination sphere: CpCo(CO) has 8 heavy (non-hydrogen) atoms. Much work is currently being done to try and reach a better degree of accuracy in treating spin-state splittings in TM complexes using ab initio methods (see [36, 65–67] for example), but this is still very difficult, so that routine calibration of DFT by ab initio methods is not really possible for the present problem.

The other approach to calibration, by comparison with experimental data, also presents difficulties. Here, the problem is in ensuring that the calculations relate correctly to the experiment, which, in this field, most frequently refers to finite-temperature *free* energy differences in the condensed phase. As well as obtaining accurate electronic energy differences, therefore, zero-point energy (zpe), thermal or enthalpic (H), entropic (S), and medium effects need to be computed or at least accounted for in some way so as to make a comparison. For TM compounds, scalar relativistic and spin-orbit coupling effects on the electronic energies are needed. The second of these in particular can be fairly large (a few kcal/mol) and is not trivial to compute. Computation of vibrational frequencies enables a reasonable approximation to the zpe, H and S terms to be computed using the standard expressions of statistical thermodynamics. However, for realistic TM complexes, this procedure is less accurate than for small closed-shell species, as one may have multiple minima, floppy (highly anharmonic) potential energy surfaces, and so on. Finally, medium effects can be important. Using a continuum solvation model, it has been found [68] that the spin-state splitting in various $Co(L)_6^{3+}$ species ($L=H_2O$ or NH_3) can differ by up to 5 kcal/mol in the gas-phase and in solution. Clearly, if such effects are general, "agreement with experiment" for a gas-phase calculation would be undesirable! To take another example, the calculations on $Mn(C_5H_5)_2$ mentioned above [40] were carried out in the gas-phase, yet the experimental spin-state splitting refers to the solid phase, in which manganocene adopts a polymeric structure, with significant inter-complex interaction, which will almost certainly affect the spin-state splitting to a significant extent. Here too, the exact correspondence between experiment and theory is thus not straightforward.

A final question concerns the interpretation of the observation that exact exchange admixture affects spin-state splittings. Why does it do this? What does this tell us about pure and hybrid functionals? What does it suggest for future functional development? The first fairly obvious point to be made is that pure density functionals do not appear to describe exchange interactions correctly in many cases. The evidence described above, whereby pure DFTs often exaggerate the stability of low-spin states compared to experiment, is so wide-ranging that it is not possible that this is due in each case to problems with interpretation of the calculations. It is in a sense no surprise that the simple, local, expressions used to evaluate exchange energy are unable to capture the complex nature of this phenomenon in all cases. However, for many main-group compounds, the problem does not appear to be a big one – fairly accurate energetics are obtained in many cases, and indeed spin-state splittings are also fairly reliable as shown by Tables 1 and 2.

From this point of view, the improvement of the description upon using hybrid functionals can be described as resulting from a fortunate cancellation of errors. However, it could be argued that this cancellation is not purely accidental, or at least that it can be partly expected on other theoretical grounds.

First, it is by now well-known that the poor cancellation of electron self-interaction in DFT can lead to a substantial error in the energy of *d* orbitals (see [69] and references therein). Self-interaction corrected DFT leads to much improved results for such systems. This has mostly been noticed in the context of periodic studies of metal oxides such as ZnO. Ziegler has pointed out [70] that the wrong electronic structure obtained for $CuCl_4^-$ with pure DFT functionals [57] is due to self-interaction errors that affect the energies of the *d* orbitals. The same author has shown recently [71] that the treatment of a number of transition states, such as that of the simple $H+H_2$ system, is much improved upon using self-interaction corrected DFT, and stated that self-interaction errors affect spin-state splittings in TM compounds [70].

Next, it is observed in other contexts that properties whose calculated values are strongly affected by the self-interaction error are improved upon mixing in progressively more HF exact exchange. The best example here is the same energy barriers for hydrogen atom abstraction reactions, whose treatment is improved with pure, but self-interaction corrected, DFTs [71]. It is well-known [37, 38] that they are poorly described by pure functionals, and that large (~50%) amounts of exact exchange are needed by functionals to generate accurate barrier heights.

Finally, in careful comparative studies of the molecular electron densities generated by HF, correlated ab initio, and pure, self-interaction corrected, and hybrid DFT calculations, Cremer et al have made a very interesting observation [72, 73]. They found that the pure DFT generated densities differed from those obtained with accurate ab initio methods in a particular way, and that both hybrid, and self-interaction corrected DFT methods, yielded densities closer to the correct ones. Based on this observation, they suggested that mixing in of exact exchange in hybrid functionals serves as a proxy for the self-interaction correction.

Based on these three points, it appears consistent with all of these observations that self-interaction errors mean that pure DFT gives a poor description of *d* orbital levels and so of spin-state splittings, and that like in other cases, exact exchange mixing can correct this error, by serving as an effective self-interaction correction. In this picture, the amount of exact exchange needed to correct the self-interaction error might well vary depending on the severity of this problem, and there would be no reason to expect systematic behaviour. Instead, one might think that accurate and reliable spin-state splittings could only be obtained for TM compounds with a new generation of functionals developed to be self-interaction error free from the start.

However, this argument certainly cannot yet prove that different amounts of HF exchange will be needed for different metal complexes. In fact, there is a considerable amount of evidence that for many compounds, an exact exchange proportion of roughly 15% gives good results. From a pragmatic point of view, therefore, the use of such modified functionals must be recommended at present, until more calculations, better calibration, and a deeper understanding of DFT performance in this area are available.

6 Conclusions

In conclusion, DFT methods must be used with caution when attempting to predict the spin-state splitting between TM compounds. Large variations, of 20 kcal/mol or more, have been observed between computed splittings using different DFT functionals. This effect is particularly strong for first-row TM systems, which are unfortunately those that most often display high-spin states! In general, "pure" DFTs tend to overestimate the stability of low-spin forms, and results obtained using such functionals must therefore be treated with great suspicion. Including HF exact exchange leads to an improved behaviour in many cases, and an optimum spin-state mixing of roughly 15% has been proposed [39, 40], which, based on current knowledge, seems to yield very good results.

Nevertheless, there are still considerable uncertainties about the generality of this observation. First, accurate calibration is difficult for TM compounds, so that it may be hard to be certain what the "correct" value is, let alone whether a given functional reproduces it! Next, a theoretical analysis suggests that the strong dependence of state-splittings on exact exchange admixture may result from the occurrence of significant self-interaction errors when using pure functionals. This analysis also invites great care in assuming that a single "best" value for exact exchange admixture can be found. Indeed, problems with self-interaction errors arise in other areas when using DFT, and can also be partially dealt with by the use of exact exchange. However, the optimum mixing in such cases is not always close to the 15% suggested to prevail for metal complexes. Clearly, DFT can be an inexpensive and useful aide for assessing spin-states of TM compounds, but much work is needed to develop new and more reliable functionals before DFT can be trusted to yield accurate spin-state energetics in the kind of "black-box" way in which B3LYP is often used nowadays.

In the meantime, computational chemists working in the area are recommended to compute spin-state splittings of importance to their particular theoretical project using several different functionals, including at least one pure functional, and one hybrid one. If large differences are found, results must be interpreted with care, and computations using hybrid functionals with ~15% exact exchange are recommended as giving the best current performance for these problems.

Acknowledgements Dr. F. R. Manby is thanked for helpful comments. JNH is an EPSRC advanced research fellow.

References

1. Poli R (1996) Chem Rev 96:2135
2. Harvey JN (2001) In: Cundari TR (ed) Computational organometallic chemistry. Marcel Dekker, New York
3. Poli R, Harvey JN (2003) Chem Soc Rev 32:1
4. Harvey JN, Poli R, Smith KM (2003) Coord Chem Rev 238:347
5. Illas F, Moreira IDR, de Graaf C, Barone V (2000) Theor Chem Acc 104:265
6. Ruiz E, Rodriguez-Fortea A, Cano J, Alvarez S, Alemany P (2003) J Comput Chem 24:982
7. Ciofini I, Daul CA (2003) Coord Chem Rev 238:187
8. Filatov M, Harris N, Shaik S (1999) Angew Chem Int Edit 38:3510
9. Jensen F (1999) Introduction to computational chemistry. Wiley, Chichester, UK
10. Siegbahn PEM (1996) In: Prigogine I, Rice SA (eds) Advances in chemical physics, Vol XCIII, Wiley, New York, p 333
11. Koch W, Holthausen MC (2001) A chemist's guide to density functional theory. Wiley-VCH, Weinheim
12. Perdew JP, Chevary JA, Vosko SH, Jackson KA, Pederson MR, Singh DJ, Fiolhais C (1992) Phys Rev B 46:6671
13. Boese AD, Doltsinis NL, Handy NC, Sprik M (2000) J Chem Phys 112:1670
14. Hammer B, Hansen LB, Norskov JK (1999) Phys Rev B 59:7413
15. Becke AD (1988) Phys Rev A 38:3098
16. Gill PMW (1996) Mol Phys 89:433
17. Perdew JP (1986) Phys Rev B 33:8822
18. Lee CT, Yang WT, Parr RG (1988) Phys Rev B 37:785
19. Becke AD (1993) J Chem Phys 98:1372
20. Becke AD (1993) J Chem Phys 98:5648
21. Bauschlicher CW, Partridge H (1995) J Chem Phys 103:1788
22. Scott AP, Platz MS, Radom L (2001) J Am Chem Soc 123:6069
23. Worthington SE, Cramer CJ (1997) J Phys Org Chem 10:755
24. Harvey JN (unpublished work)
25. Baerends EJ, Branchadell V, Sodupe M (1997) Chem Phys Lett 265:481
26. Mineva T, Goursot A, Daul C (2001) Chem Phys Lett 350:147
27. Weinert M, Watson RE, Fernando GW (2002) Phys Rev A 66:032508
28. Markovits A, Paniagua JC, Lopez N, Minot C, Illas F (2003) Phys Rev B 67:115417
29. Daniel C, Benard M, Dedieu A, Wiest R, Veillard A (1984) J Phys Chem 88:4805
30. Barnes LA, Rosi M, Bauschlicher CW (1991) J Chem Phys 94:2031
31. Delley B, Wrinn M, Luthi HP (1994) J Chem Phys 100:5785
32. Wang WH, Weitz E (1997) J Phys Chem A 101:2358
33. Decker SA, Klobukowski M (1998) J Am Chem Soc 120:9342
34. Gonzalez-Blanco O, Branchadell V (1999) J Chem Phys 110:778
35. Matveev A, Staufer M, Mayer M, Rosch N (1999) Int J Quantum Chem 75:863
36. Harvey JN, Aschi M (2003) Faraday Discuss 124:129
37. Kang JK, Musgrave CB (2001) J Chem Phys 115:11040
38. Lynch BJ, Fast PL, Harris M, Truhlar DG (2000) J Phys Chem A 104:4811
39. Reiher M, Salomon O, Hess BA (2001) Theor Chem Acc 107:48
40. Salomon O, Reiher M, Hess BA (2002) J Chem Phys 117:4729
41. Deng LQ, Margl P, Ziegler T (1999) J Am Chem Soc 121:6479
42. Khoroshun DV, Musaev DG, Vreven T, Morokuma K (2001) Organometallics 20:2007
43. Margl P, Deng LQ, Ziegler T (1999) Organometallics 18:5701

44. Ghosh A, Taylor PR (2003) Curr Opin Chem Biol 7:113
45. Scherlis DA, Estrin DA (2002) Int J Quantum Chem 87:158
46. Paulsen H, Duelund L, Winkler H, Toftlund H, Trautwein HX (2001) Inorg Chem 40:2201
47. Reiher M (2002) Inorg Chem 41:6928
48. Chen GJ, Espinosa-Perez G, Zentella-Dehesa A, Silaghi-Dumitrescu I, Lara-Ochoa F (2000) Inorg Chem 39:3440
49. Chen GJ, Liu RZ, Silaghi-Dumitrescu I, Espinosa-Perez G, Zentella-Dehesa A, Lara-Ochoa F (2001) Int J Quantum Chem 83:60
50. Brandt P, Norrby PO, Daly AM, Gilheany DG (2002) Chem-Eur J 8:4299
51. Linde C, Akermark B, Norrby PO, Svensson M (1999) J Am Chem Soc 121:5083
52. Abashkin YG, Collins JR, Burt SK (2001) Inorg Chem 40:4040
53. Khavrutskii IV, Musaev DG, Morokuma K (2003) Inorg Chem 42:2606
54. Cavallo L, Jacobsen H (2003) J Phys Chem A 107:5466
55. Jensen VR, Poli R (2003) J Phys Chem A 107:1424
56. Ball DM, Buda C, Gillespie AM, White DP, Cundari TR (2002) Inorg Chem 41:152
57. Szilagyi RK, Metz M, Solomon EI (2002) J Phys Chem A 106:2994
58. Wilson PJ, Tozer DJ (2002) J Chem Phys 116:10139
59. Siegbahn PEM (1996) J Am Chem Soc 118:1487
60. Carréon-Macedo J-L, Harvey JN (2004) J Am Chem Soc 126:5789
61. Lee TJ, Taylor PR (1989) Int J Quantum Chem, Quantum Chem Symp 23:199
62. Gagliardi L, Ziegler, Neville S, Garner, Green, Harvey, Eisenstein, Bursten, Kaltsoyannis, Macgregor, Perutz (2003) Faraday Discuss 124:145
63. Curtiss LA, Raghavachari K, Redfern PC, Rassolov V, Pople JA (1998) J Chem Phys 109:7764
64. Curtiss LA, Raghavachari K (2002) Theor Chem Acc 108:61
65. Pierloot K (2003) Mol Phys 101:2083
66. Roos BO, Widmark PO, Gagliardi L (2003) Faraday Discuss 124:57
67. Jensen VR, Thiel W (2001) Organometallics 20:4852
68. Schmiedekamp AM, Ryan MD, Deeth RJ (2002) Inorg Chem 41:5733
69. Meyer B, Marx D (2003) Phys Rev B 67:035403
70. Ziegler T (2003) Faraday Discuss 124:287
71. Patchkovskii S, Ziegler T (2002) J Chem Phys 116:7806
72. Cremer D (2001) Mol Phys 99:1899
73. Polo V, Kraka E, Cremer D (2002) Mol Phys 100:1771

Author Index Volumes 101–112

Author Index Vols. 1–100 see Vol. 100

The volume numbers are printed in italics

Subject Index